# STATE OF THE World's Forests 2007

FOOD AND AGRICULTURE ORGANIZATION OF THE UNITED NATIONS

Rome, 2007

Produced by the
**Electronic Publishing Policy and Support Branch**
**Communication Division**
**FAO**



ISBN 978-92-5-105586-1

# Contents

## Annex

# Foreword

**FAO'S BIENNIAL** *State of the World's Forests* series offers a global perspective on the forest sector, including its environmental, economic and social dimensions.

Two years is a short time in the life of a forest, and in most international processes, too. So what is new in forestry since the last edition of *State of the World's Forests*? First, the release of the results of the Global Forest Resources Assessment 2005 (FRA 2005) has provided new information, more comprehensive than ever, for evaluating the state of the forests. The Kyoto Protocol has come into force, with significant implications for forestry. New initiatives have been developed, such as networks for information sharing and action on forest invasive species, efforts to link national forest programmes and poverty reduction strategies, and the development of guiding principles on planted forests and fire management. Even the structure and look of *State of the World's Forests* are new.

This seventh edition examines progress towards sustainable forest management. The analysis reveals that some countries and some regions are making more progress than others. Most countries in Europe and North America have succeeded in reversing centuries of deforestation and are now showing a net increase in forest area. Most developing countries, especially those in tropical areas, continue to experience high rates of deforestation and forest degradation. The countries that face the most serious challenges in achieving sustainable forest management are, by and large, the countries with the highest rates of poverty and civil conflict.

Part I reviews progress at the regional level. This section was developed from six regional reports prepared for discussion in 2006 by FAO's six regional forestry commissions. Each regional summary is structured according to the seven thematic elements of sustainable forest management that were agreed by international fora as a framework for sustainable forest management. The regional reports synthesize the most current information available, including data gathered by FAO for FRA 2005 (which was, in turn, based on country reports submitted to FAO and the contributions of over 800 people, including 172 national correspondents), the FAOSTAT online database (compiling economic information provided by countries) and recent FAO regional forestry sector outlook studies, as well as input from FAO partners.

Part II presents selected issues in the forest sector, addressing the latest developments in 18 topics of interest to forestry. In a few pages each, FAO specialists present the state of knowledge or latest activities on themes ranging from climate change and desertification to wildlife management and wood energy.

FAO is pleased to publish *State of the World's Forests 2007* and trusts that readers will find it stimulating and informative.

**Jan Heino**
Assistant Director-General
FAO Forestry Department

# Acknowledgements

**THE COMPILATION** of *State of the World's Forests 2007* was coordinated by D. Kneeland. Special thanks go to L. Ball, who edited the publication.

The following FAO staff wrote or reviewed sections of the report or assisted with tables, maps or graphics: G. Allard, A. Branthomme, J. Carle, C. Carneiro, F. Castañeda, P. Durst, M. Gauthier, O. Hashiramoto, T. Hofer, P. Holmgren, O. Jonsson, W. Killmann, P. Kone, J.P. Koyo, A. Lebedys, J. Lorbach, M. Malagnoux, E. Mansur, L.G. Marklund, M. Martin, R. McConnell, E. Muller, C.T.S. Nair, A. Perlis, J.A. Prado, D. Reeb, D. Schoene, M. Trossero, T. Vahanen, P. Vuorinen, M. Wilkie and D. Williamson – benefiting from the wealth of information provided by international partners.

FAO thanks the International Tropical Timber Organization for its contribution summarizing the *Status of tropical forest management 2005*.

A. Perlis, L. Frezza and the staff of the FAO Electronic Publishing Policy and Support Branch provided editorial and production support.

# Acronyms

| | |
|---|---|
| APFISN | Asia–Pacific Forest Invasive Species Network |
| ARC | Alliance of Religions and Conservation |
| C&I | criteria and indicators |
| CBD | Convention on Biological Diversity |
| CDM | Clean Development Mechanism |
| CIFOR | Center for International Forestry Research |
| CITES | Convention on International Trade in Endangered Species of Wild Fauna and Flora |
| COMIFAC | Conference of Ministers in Charge of Forests in Central Africa |
| COFO | FAO Committee on Forestry |
| CPF | Collaborative Partnership on Forests |
| FISNA | Forest Invasive Species Network for Africa |
| FRA | Global Forest Resources Assessment |
| GDP | gross domestic product |
| GISP | Global Invasive Species Programme |
| ICIMOD | International Centre for Integrated Mountain Development |
| ICRAF | World Agroforestry Centre |
| IEA | International Energy Agency |
| IPPC | International Plant Protection Convention |
| ISPM | International Standard for Phytosanitary Measures |
| ITTA | International Tropical Timber Agreement |
| ITTO | International Tropical Timber Organization |
| IUCN | World Conservation Union |
| IUFRO | International Union of Forest Research Organizations |
| IYDD | International Year of Deserts and Desertification |
| JRC | Joint Research Centre of the European Commission |
| MCPFE | Ministerial Conference on the Protection of Forests in Europe |
| NEPAD | New Partnership for Africa's Development |
| NGO | non-governmental organization |
| NWFP | non-wood forest product |
| OECD | Organisation for Economic Co-operation and Development |
| PROFOR | Program on Forests |
| REDLACH | Red Latinoamericana de Cooperación Técnica en Manejo de Cuencas Hidrográficas |
| REDPARQUES | Red Latinoamericana de Cooperación Técnica en Parques Nacionales, otras Áreas Protegidas, Flora y Fauna Silvestres |
| SADC | Southern African Development Community |
| UNCCD | United Nations Convention to Combat Desertification |
| UNCED | United Nations Conference on Environment and Development |
| UNDP | United Nations Development Programme |
| UNECE | United Nations Economic Commission for Europe |
| UNEP | United Nations Environment Programme |
| UNEP-WCMC | World Conservation Monitoring Centre (UNEP) |
| UNFCCC | United Nations Framework Convention on Climate Change |
| UNFF | United Nations Forum on Forests |
| USDA | United States Department of Agriculture |
| WWF | World Wide Fund for Nature |

# Summary

**THIS SEVENTH BIENNIAL** issue of *State of the World's Forests* considers progress towards sustainable forest management at the regional and global levels. The overall conclusion is that progress is being made, but is very uneven. Some regions, notably those including developed countries and having temperate climates, have made significant progress; institutions are strong, and forest area is stable or increasing. Other regions, especially those with developing economies and tropical ecosystems, continue to lose forest area, while lacking adequate institutions to reverse this trend. However, even in regions that are losing forest area, there are a number of positive trends on which to build.

The biggest limitation for evaluating progress is weak data. Relatively few countries have had recent or comprehensive forest inventories. With many partners, FAO is assisting countries in carrying out national forest assessments and strengthening forestry institutions, but progress is slow, owing in part to the scarcity of financial resources.

## PROGRESS TOWARDS SUSTAINABLE FOREST MANAGEMENT

### Africa

During the 15-year period from 1990 to 2005, Africa lost more than 9 percent of its forest area. In a typical year, Africa accounts for more than half of the global forest area damaged by wildfire. Deforestation and uncontrolled forest fires are especially severe in countries suffering from war or other civil conflict. Most forests in Africa are owned by national governments, and the national forest agencies in many countries lack the financial resources required to manage the forest resources sustainably.

But the picture is not all gloomy. Forests are obtaining political support and commitment at the highest levels in Africa. For example, the Conference of Ministers in Charge of Forests in Central Africa (COMIFAC) ranks among the world's most effective examples of regional collaboration among countries to address serious environmental issues. During the period from 2000 to 2005, African countries designated over 3.5 million hectares of forest to be managed primarily for conservation of biological diversity, raising the total to almost 70 million hectares. A majority of countries in the region have adopted new forest policies and forest laws, and efforts are being made in many countries to improve law enforcement and governance.

### Asia and the Pacific

The good news for the Asia and the Pacific region is that net forest area increased between 2000 and 2005, reversing the downward trend of the preceding decades. However, the increase was limited to East Asia, where a large investment in forest plantations in China was enough to offset high rates of deforestation in other areas. The net loss of forest area actually accelerated in Southeast Asia, and, in South Asia, a small increase in forest area during the 1990s was followed by a small decrease between 2000 and 2005.

However, there are a number of positive trends in the region that support an optimistic view of the future. Rapid economic growth in the two largest countries, China and India, may help to create the conditions for sustainable forest management. Economic development appears to be a necessary condition for deforestation to cease. Employment in the forest sector and trade of forest products are both increasing. Forest institutions in the region are getting stronger in a number of countries, and the long-term trend towards more participatory decision-making continues.

On the other hand, economic development creates new problems. There is evidence that illegal logging is increasing in some countries in the region in response to the high demand for log imports in other countries with rapidly growing economies. Forest disturbances by pests and diseases pose a significant threat to forests, and this is an important issue for new forest plantations. Forest fires may increase in severity if the global climate continues to become warmer and more variable.

### Europe

It is tempting to conclude that Europe has achieved sustainable forest management. Forest area is increasing in most countries, and the positive trends exceed the negative. Forest institutions are strong, and changes in forest policies and institutions are largely positive. The Ministerial Conference on the Protection of Forests in Europe (MCPFE) is the strongest regional political mechanism to address forest issues in the world.

However, there are a number of areas of concern. Employment in the forest sector continues to decline, and the forest sector's contribution to the economy is declining in comparison with that of many other sectors. Forests remain vulnerable to disturbances that are likely to increase if the global climate continues to change as many experts predict. Countries with economies in transition are striving to improve support and guidance to owners of newly privatized forests.

### Latin America and the Caribbean

Latin America and the Caribbean joins Africa as the two regions that are losing their forests at the highest rates. The annual net rate of loss between 2000 and 2005 (0.51 percent) was higher than that of the 1990s (0.46 percent). Countries in the region are fighting an uphill battle to retain their primary forests, but they are making considerable efforts, including an annual increase of over 2 percent in the area of forest designated primarily for conservation of biological diversity.

Regional and subregional cooperation to address forest issues is gaining strength. Latin American countries have formed networks to fight forest fires, to increase the effectiveness of protected area management and to improve watershed management. Employment and trade in the forest sector are increasing, and institutions are getting stronger. Several countries in the region are among the global leaders in innovative approaches to forest management, such as payments for environmental services.

### Near East

Largely because of the arid climate, the forest sector in the Near East region represents a small part of the economy. Countries in the region rely heavily on forest product imports. However, there have been significant investments in forest plantations in recent years. In comparison with other regions, trees outside forests are important for both the environment and the economy.

Countries that are experiencing conflict are having the most difficulty managing their forests and controlling deforestation. Several countries have been successful in using incentives to promote good forest management. Despite the problems and limitations faced by countries in the region, progress is being made to develop strategies and implement programmes that effectively address local conditions.

### North America

The North America region includes only three countries: Canada, Mexico and the United States of America; but all three have significant forest resources and highly developed forest institutions. Net forest area is stable in Canada and the United States of America. It is declining in Mexico, but the rate of decrease is slowing and is much less than the rate of forest loss in Central America. North American forests account for 17 percent of the world's forest area and 40 percent of the world's wood removals, suggesting that the region's forests are relatively productive and the commercial sector is relatively advanced.

However, while the region's forest resources remain abundant, the forest sector's contribution to the regional economy is declining. Employment in the forest sector is fairly flat, and the region as a whole has fallen from a major net exporter of forest products to a major net importer. This reversal is mainly a result of the sharp decline in the forest products trade balance of the United States of America, whose forest products exports exceeded its imports in the early 1990s but now trail them by half.

### The global view

Forestry makes a valuable contribution to sustainable development in all parts of the world, but progress towards sustainable forest management has been uneven. The world has just under 4 billion hectares of forest, covering about 30 percent of the world's land area. From 1990 to 2005, the world lost 3 percent of its total forest area, an average decrease of some 0.2 percent per year.

Many countries have demonstrated the political will to improve management of their forests by revising forest policies and legislation and strengthening forestry institutions. Most countries manage forests for multiple uses, and increasing attention is being paid to the conservation of soil, water, biological diversity and other environmental values. However, the continuing decline in primary forests in most tropical countries is a matter of serious concern.

The world is faced with an increasingly complex challenge: is it possible to achieve sustainable forest management and to achieve equitable economic progress at the same time?

# Summary

## SELECTED ISSUES IN THE FOREST SECTOR

### Climate change

Evidence is mounting that forests will be profoundly affected by climate change, such as the increasing damage to forest health caused by proliferation of fire, pests and diseases. At the same time, new investments in forests to mitigate climate change lag behind the optimistic expectations of many following the entry into force of the Kyoto Protocol in 2005.

### Desertification

While all dry regions of the world are affected by land degradation, the world's highest rate of desertification is taking place in sub-Saharan Africa, where agricultural productivity is declining at a rate of almost 1 percent per year. Effective action to combat desertification requires an integrated approach, including investments in afforestation.

### Forest landscape restoration

There is an emerging global consensus that forests need to be managed from a broad multidisciplinary perspective. The forest landscape restoration concept emphasizes the importance of bringing people together to develop practices that restore the balance among the ecological, social, cultural and economic benefits of forests and trees, within the broader pattern of land uses.

### Forestry and poverty reduction

Many countries are shifting strategies to address more effectively the need for the forest sector to contribute to poverty reduction, starting with the recognition of forest benefits, which are systematically undervalued in almost all countries.

### Forestry sector outlook

Global forest sector outlook studies provide countries with critical information to manage their forests. Each regional study involves broad stakeholder participation in comprehensive reviews of the socio-economic changes affecting the region and the global economy. Five regional studies have been completed, and a new study is under way to extend the earlier outlook for Asia and the Pacific from 2010 to 2020.

### Forest tenure

Secure forest tenure and access to forest resources are prerequisites to sustainable forest management. At the global level, 84 percent of forest lands and 90 percent of other wooded land are publicly owned. The area of forests owned and administered by communities doubled from 1985 to 2000, reaching 22 percent in developing countries. The transfer of forest management and user rights needs to be (but often is not) accompanied by adequate security of tenure and the capacity to manage these resources.

### Harvesting

Good harvesting practices can be profitable and can significantly reduce the environmental impacts of forest harvesting. However, inappropriate harvesting methods are still used widely throughout the tropics. Illegal logging and lack of awareness are among the main reasons. A number of regional and national codes of practice have been adopted, but implementation remains slow.

### Invasive species

Awareness of the problem of forest invasive species has become heightened in recent years. Land-use changes, forest management activities, tourism and trade facilitate potentially harmful introductions. Numerous international and regional programmes and instruments, binding and non-binding, have been developed to address the problem, of which a number have direct or indirect implications for forests and the forest sector.

### Monitoring, assessment and reporting

In recent years, significant progress has been made in monitoring, assessment and reporting on forests. Criteria and indicators are used to monitor progress towards sustainable forest management, especially at the national level. New tools are being developed to improve monitoring, assessment and reporting as relates to international commitments, but a heavy reporting burden remains, with new obligations in many fora. A major future challenge will be to mobilize the resources to invest in basic information and knowledge management to ensure that forest-related decisions are based on sound data.

### Mountain development

Since the International Year of Mountains in 2002, mountain issues have gained increasing attention. The membership and visibility of the Mountain Partnership are rapidly expanding (over 130 government, private and non-governmental organization [NGO] members). This growth underscores the need for improved approaches and increased investments in the livelihoods of the more than 700 million people who live in mountain regions.

### Payment for environmental services

Conventional wisdom suggests that forest benefits are undervalued by markets; the question is what to do about it. Some countries have developed payment-for-environmental-services schemes as a way to reward forest owners for the production of non-market benefits. As a prerequisite to such schemes, countries may wish to ensure that charges and taxes on forest producers are effectively established and collected and that the proceeds are reinvested in the forest.

### Planted forests

Planted forests continue to expand, and their contribution to global wood production is approaching 50 percent of the total. New information gathered in 2005 on trends in planted forests indicates that the areas of forests planted for production and of those planted for protective purposes are both steadily increasing in all regions except Africa.

### Trade in forest products

Forest products trade continues to expand. In 2004, trade in industrial roundwood was 120 million cubic metres, or about 7 percent of global production, with a value of US$327 billion. Each of these figures established records for trade in the forest sector. As trade has boomed, a number of developed countries have adopted public procurement policies to promote the use of legally or sustainably produced forest products.

### Urban forestry

The urbanization of society poses immense challenges for forestry and has new impacts on forests. Urban forestry is increasingly recognized as an important economic and social component of effective urban planning.

### Voluntary tools for sustainable forest management

A continuum of tools for advancing sustainable forest management is available to policy-makers and forest managers, ranging in approach from incentive-based and voluntary to legally binding, and in scope from local to global. These include criteria and indicators, certification, codes and guidelines, and initiatives to promote forest law enforcement.

### Water

Several interesting recent studies challenge conventional views on the relationship between forests and water. More trees may not always result in more water for humans, and fewer trees may not result in catastrophic floods.

### Wildlife management

Several important wildlife species experienced severe declines in the past century. Unsustainable hunting and trading in wildlife and wildlife products, and conflicts between humans and wildlife (including injuries or death to both, as well as damage to property and crops), are persistent problems. A challenge for policy-makers is to balance conservation of wildlife resources with the livelihood requirements of local populations.

### Wood energy

As the price of oil soars, alternative energy sources are receiving increased attention. In Africa, wood is by far the major source of energy; in other regions, wood may become a major energy source in the future as it was in the past.

PART 1

# Progress towards sustainable forest management

S. Verver

Part I examines progress towards sustainable forest management region by region. Broadly speaking, sustainable forest management refers to the use and conservation of forests for the benefit of present and future generations. It is clearly an issue of widespread interest. A Google™ search for "sustainable forest management" produces 25 million results.

The concept of sustainable forest management gained momentum during the 1990s when forest issues were debated within the wider framework of sustainable development, which has several broad dimensions: environmental, economic, social and cultural. A number of countries have sponsored processes to identify criteria and indicators (C&I) for sustainable forest management. Building upon C&I processes, intergovernmental processes such as the United Nations Forum on Forests (UNFF) have identified seven thematic elements (Box 1) as a framework for monitoring, assessing and reporting on progress towards sustainable forest management:

- extent of forest resources
- biological diversity
- forest health and vitality
- productive functions of forest resources
- protective functions of forest resources
- socio-economic functions
- legal, policy and institutional framework.

*State of the World's Forests 2007* uses these seven elements as a framework for discussing progress towards sustainable forest management.

The first six elements were used as the framework for the most recent *Global Forest Resources Assessment* (FRA 2005) (FAO, 2006a). Unless stated otherwise, data discussed in Part I are taken from FRA 2005. Part I also draws on economic statistics published in the FAOSTAT online database (FAO, 2006b) and on information gathered for forestry sector outlook studies and national forest programme updates. All of these sources rely heavily on information provided by national correspondents. Hence, the present text is essentially based on information provided by countries.

In addition, a number of other sources were used to validate data, including official national Web sites and reports, remote sensing studies and expert assessments. Regional reports were discussed at the 2006 sessions of the FAO regional forestry commissions, whose comments have been incorporated.

**BOX 1** Thematic elements of sustainable forest management

1. **Extent of forest resources.** This theme reflects the importance of adequate forest cover and stocking, including trees outside forests, to support the social, economic and environmental dimensions of forestry; to reduce deforestation; and to restore and rehabilitate degraded forest landscapes. The existence and extent of specific forest types are important as a basis for conservation efforts. The theme also includes the important function of forests and trees outside forests to store carbon and thereby contribute to moderating the global climate.
2. **Biological diversity.** This theme concerns the conservation and management of biological diversity at ecosystem (landscape), species and genetic levels. Such conservation, including the protection of areas with fragile ecosystems, ensures that diversity of life is maintained, and provides opportunities to develop new products in the future, including medicines. Genetic improvement is also a means of increasing forest productivity, for example to ensure high wood production levels in intensively managed forests.
3. **Forest health and vitality.** Forests need to be managed so that the risks and impacts of unwanted disturbances are minimized, including wildfires, airborne pollution, storm felling, invasive species, pests and diseases. Such disturbances may have an impact on the social and economic, as well as environmental, dimensions of forestry.
4. **Productive functions of forest resources.** Forests and trees outside forests provide a wide range of wood and non-wood forest products. This theme reflects the importance of maintaining an ample and valuable supply of primary forest products while ensuring that production and harvesting are sustainable and do not compromise the management options of future generations.
5. **Protective functions of forest resources.** Forests and trees outside forests contribute to moderating soil, hydrological and aquatic systems, maintaining clean water (including healthy fish populations) and reducing the risks and impacts of floods, avalanches, erosion and drought. Protective functions of forest resources also contribute to ecosystem conservation efforts and provide benefits to agriculture and rural livelihoods.
6. **Socio-economic functions.** Forest resources contribute to the overall economy in many ways such as through employment, values generated through processing and marketing of forest products, and energy, trade and investment in the forest sector. They also host and protect sites and landscapes of high cultural, spiritual or recreational value. This theme thus includes aspects of land tenure, indigenous and community management systems, and traditional knowledge.
7. **Legal, policy and institutional framework.** Legal, policy and institutional arrangements – including participatory decision-making, governance and law enforcement, and monitoring and assessment of progress – are necessary to support the above six themes. This theme also encompasses broader societal aspects, including fair and equitable use of forest resources, scientific research and education, infrastructure arrangements to support the forest sector, transfer of technology, capacity-building, and public information and communication.

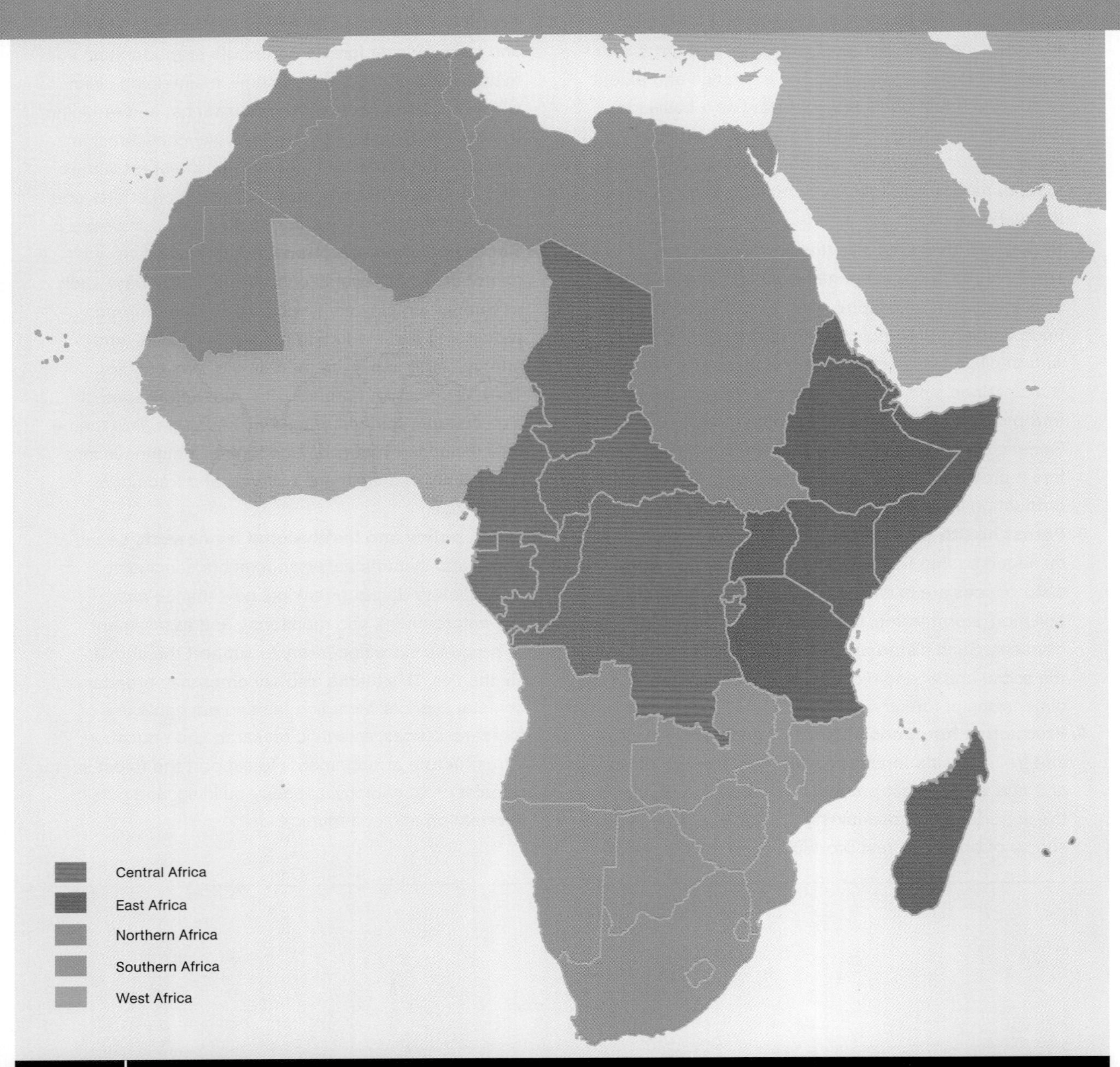


**FIGURE 1** Subregional breakdown used in this report

**Central Africa:** Burundi, Cameroon, Central African Republic, Chad, Congo, Democratic Republic of the Congo, Equatorial Guinea, Gabon, Rwanda, Saint Helena, Sao Tome and Principe

**East Africa:** British Indian Ocean Territory, Comoros, Djibouti, Eritrea, Ethiopia, Kenya, Madagascar, Mauritius, Mayotte, Réunion, Seychelles, Somalia, Uganda, United Republic of Tanzania

**Northern Africa:** Algeria, Egypt, Libyan Arab Jamahiriya, Mauritania, Morocco, Sudan, Tunisia

**Southern Africa:** Angola, Botswana, Lesotho, Malawi, Mozambique, Namibia, South Africa, Swaziland, Zambia, Zimbabwe

**West Africa:** Benin, Burkina Faso, Cape Verde, Côte d'Ivoire, Gambia, Ghana, Guinea, Guinea-Bissau, Liberia, Mali, Niger, Nigeria, Senegal, Sierra Leone, Togo

**NOTE:** For consistency, this report uses the same subregional groups that were used in the *Forestry Outlook Study for Africa* (FAO, 2003).

# Africa

## EXTENT OF FOREST RESOURCES

The estimated forest area for Africa in 2005 is 635 million hectares (Figure 2), accounting for about 16 percent of global forest area. Net annual forest loss is about 4 million hectares for the period 2000–2005 (Table 1). This amounts to almost 55 percent of the global reduction in forest area. However, the reported forest cover is distributed unevenly among the different subregions and countries.

A significant share of net forest loss is reported from those countries with the greatest extent of forests. For example, Angola, the United Republic of Tanzania and Zambia together account for a majority of the forest loss in East and Southern Africa (Figure 3). Available information also indicates a high rate of forest loss in Zimbabwe, estimated at 1.7 percent per year, far above the average of 0.7 percent for all Southern Africa. In Northern Africa, the Sudan alone accounts for most of the forest cover and for 60 percent of the forest reduction. In West and Central Africa, Cameroon, the Democratic Republic of the Congo and Nigeria together account for most of the loss.

TABLE 1
**Extent and change of forest area**

| Subregion | Area (1 000 ha) | | | Annual change (1 000 ha) | | Annual change rate (%) | |
|---|---|---|---|---|---|---|---|
| | 1990 | 2000 | 2005 | 1990–2000 | 2000–2005 | 1990–2000 | 2000–2005 |
| Central Africa | 248 538 | 239 433 | 236 070 | –910 | –673 | –0.37 | –0.28 |
| East Africa | 88 974 | 80 965 | 77 109 | –801 | –771 | –0.94 | –0.97 |
| Northern Africa | 84 790 | 79 526 | 76 805 | –526 | –544 | –0.64 | –0.69 |
| Southern Africa | 188 402 | 176 884 | 171 116 | –1 152 | –1 154 | –0.63 | –0.66 |
| West Africa | 88 656 | 78 805 | 74 312 | –985 | –899 | –1.17 | –1.17 |
| **Total Africa** | **699 361** | **655 613** | **635 412** | **–4 375** | **–4 040** | **–0.64** | **–0.62** |
| **World** | **4 077 291** | **3 988 610** | **3 952 025** | **–8 868** | **–7 317** | **–0.22** | **–0.18** |

**FIGURE 2** Extent of forest resources

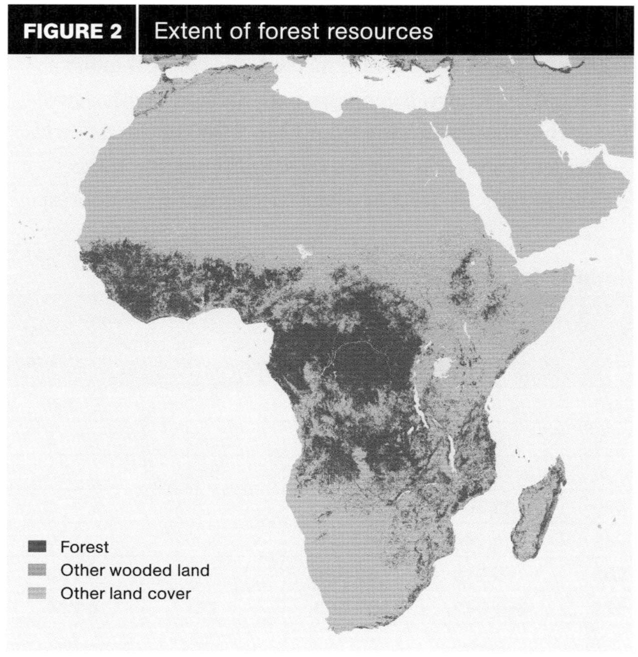


**FIGURE 3** Forest change rates by country or area, 2000–2005

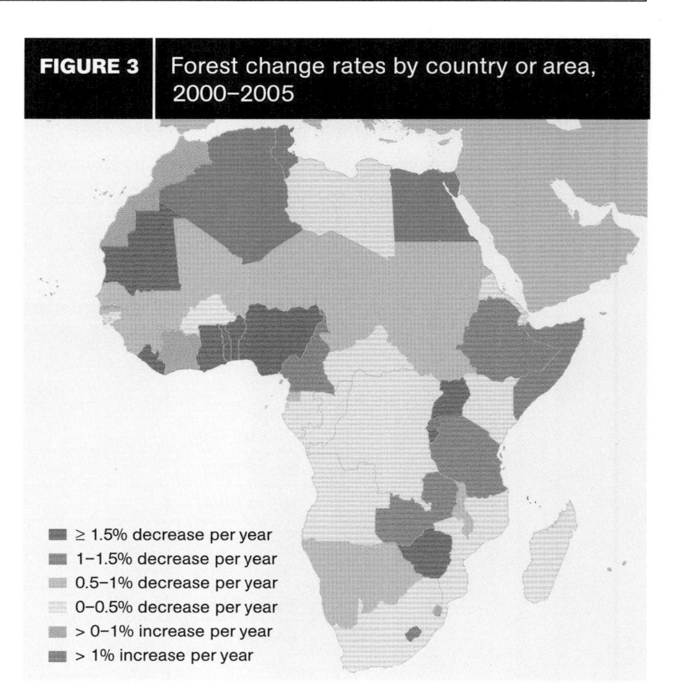


**SOURCE:** FAO, 2001a.

TABLE 2
**Area of forest plantations**

| Subregion | Area (1 000 ha) | | | Annual change (1 000 ha) | |
|---|---|---|---|---|---|
| | 1990 | 2000 | 2005 | 1990–2000 | 2000–2005 |
| Central Africa | 348 | 388 | 526 | 4 | 28 |
| East Africa | 1 246 | 1 233 | 1 230 | −1 | −1 |
| Northern Africa | 7 696 | 7 513 | 7 503 | −18 | −2 |
| Southern Africa | 1 867 | 2 060 | 2 150 | 19 | 18 |
| West Africa | 900 | 1 337 | 1 677 | 44 | 68 |
| **Total Africa** | **12 057** | **12 532** | **13 085** | **48** | **111** |
| **World** | **101 234** | **125 525** | **139 466** | **2 424** | **2 788** |

Africa also has more than 400 million hectares of "other wooded land", with scattered trees but not enough to be defined as "forest". Data on the extent and growing stock of other wooded land are weak, but the extent continues to decline.

Africa's total area of forest plantations – a subset of planted forests defined as those consisting primarily of introduced species – is about 13.0 million hectares (Table 2). Approximately 2.4 million hectares (18 percent) of forest plantations are planted for protective purposes; the remainder are planted to produce wood, particularly industrial roundwood and fuelwood. Most forest plantations are in Northern Africa, which is dependent on plantations because of the scarcity of natural forests. Southern Africa has developed a globally competitive forest industry almost entirely based on planted forests.

Since 1990, forest cover in Africa has been declining at one of the highest rates in the world (together with Latin America and the Caribbean). However, the rate of loss has shown signs of declining slightly in the past five years. Unfortunately, there are only a few countries in which forest cover is increasing or marginally improving, and most of these are the "low forest cover" countries of Northern Africa in which substantial efforts have been made to establish planted forests. Improvement has been reported in Rwanda and Swaziland as well, also largely resulting from increased planting.

## BIOLOGICAL DIVERSITY

There is evidence of an overall decline in the area of primary forests in the region, but some of the most important forested countries were not able to report on this parameter, especially in Central Africa. Thus, it is not possible to make a definitive statement regarding the magnitude of this trend.

Forty-three countries, accounting for some 70 percent of the forest area in Africa, provided information on area of forest designated for biodiversity conservation for the three reference years. In these countries, a total of about 69.5 million hectares of forests, accounting for about 16 percent of the forest area, are designated primarily for conservation (Table 3).

Although the area so designated declined in some countries, at the regional level there has been a substantial increase, especially during 2000–2005.

Forest composition, the number of native forest species and the existence (or absence) of threatened and endangered species are other indicators of biodiversity. However, with only 16 countries reporting on these variables, a clear indication of the state of biodiversity is not available. Forest composition and the preponderance of species differ widely within Africa. As would be expected, the tropical moist forests in the Congo Basin have high diversity, with native forest tree species varying from 12 to 5 000 in the reporting countries (Figure 4). The ten most common tree species represent only 22 percent of the species in a typical forest

TABLE 3
**Area of forest designated primarily for conservation**

| Subregion | Area (1 000 ha) | | | Annual change (1 000 ha) | |
|---|---|---|---|---|---|
| | 1990 | 2000 | 2005 | 1990–2000 | 2000–2005 |
| Central Africa | 26 497 | 26 375 | 30 388 | −12 | 803 |
| East Africa | 2 934 | 2 882 | 2 818 | −5 | −13 |
| Northern Africa | 9 773 | 9 051 | 8 687 | −72 | −73 |
| Southern Africa | 12 360 | 12 360 | 12 360 | 0 | 0 |
| West Africa | 15 239 | 15 244 | 15 275 | 0 | 6 |
| **Total Africa** | **66 803** | **65 912** | **69 528** | **-89** | **723** |
| **World** | **298 424** | **361 092** | **394 283** | **6 267** | **6 638** |

**NOTE:** Fewer than 50 percent of the countries in Central Africa were able to provide data on this parameter for all three years.

FIGURE 4 Number of native tree species

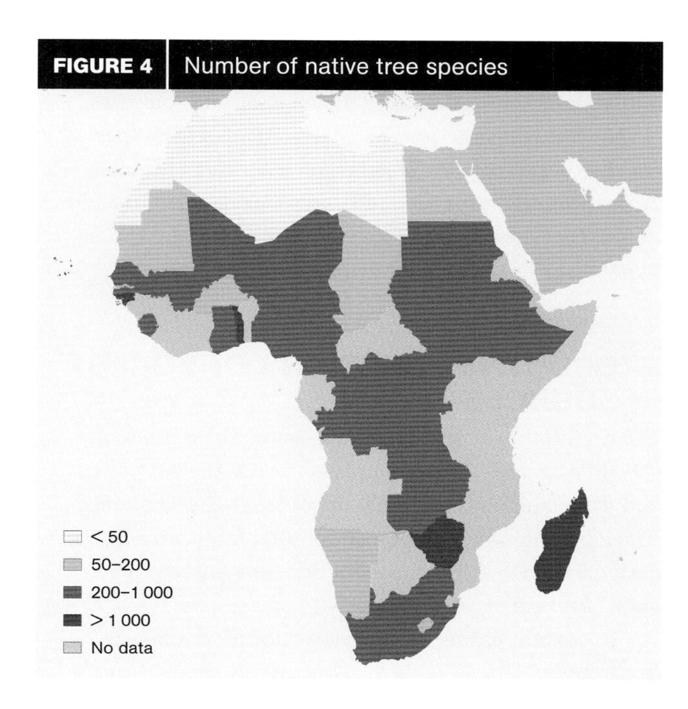


unit. In a temperate or boreal forest, the most common ten species account for over 50 percent.

Country statistics for the number of threatened tree species are more reliable, owing to regular monitoring and reporting for the *IUCN Red List of Threatened Species* (IUCN [World Conservation Union], 2000 and 2004). On average, each African country lists about 7 percent of its native tree species as critically endangered, endangered or vulnerable.

## FOREST HEALTH AND VITALITY

In Africa, as in several other regions, it is difficult to analyse trends in forest health because of the scarcity of information. Only 14 countries of 58 provided information on trends in forest fires over two time periods, accounting for 19 percent of the total forests in Africa. However, the Joint Research Centre of the European Commission (JRC) carried out a remote sensing study of wildland fires in Africa (including, but not limited to, forest fires) (JRC, 2000). The study concluded that Africa accounted for 64 percent of the global area burned by wildland fires in 2000, when 230 million hectares were burned, accounting for 7.7 percent of the total land area of the continent. A follow-up study in 2004 revealed similar results.

As reported to the 2005 FAO Regional Conference for Africa (FAO, 2006c), two areas of particularly high fire frequency stand out: one is northern Angola and the southern Democratic Republic of the Congo, and the other southern Sudan and the Central African Republic (Figure 5). These areas were once mostly tropical forest, but today the vegetation is a mosaic of grassland and remnant tropical forest patches, interspersed with the fields and settlements of both sedentary and shifting cultivators. Most of the deforestation of this zone took place decades ago, and while fire is undoubtedly preventing forest regeneration, its prevalence is symptomatic of the past conversion of forest to grassland, rather than being the direct, current cause of forest loss.

The number of fires and the area burned vary considerably from year to year, often in synchrony with the El Niño Southern Oscillation (ENSO) and associated extreme weather phenomena. Rainfall, biomass production and ENSO are particularly strongly linked in Southern Africa. A comparison of the extent of burning there in 1992, when the region experienced a severe drought, and in 2000, following a season of above-average rainfall, showed much earlier and more extensive burning in 2000. Good rains produce more biomass, and therefore more fuel for fire during the dry season, unless the additional production is consumed by livestock or wildlife. This link between rainfall and biomass production means that regional, seasonal climate forecasts can be used to anticipate the likely vegetation biomass conditions in the coming season and to assess the level of fire risk.

Damage from wildfire is a significant threat to sustainable forest management in Africa. Long-term data are not sufficient to conclude whether the area affected by forest fires is increasing or declining. Moreover, it is difficult to generalize whether a decline in fire incidence is a positive development or not. In several ecosystems, fire is an integral part of the natural ecosystem processes.

As regards the incidence of pests and diseases, comparable data over a period of time were provided in

FIGURE 5 Extent of burning, 2004/2005

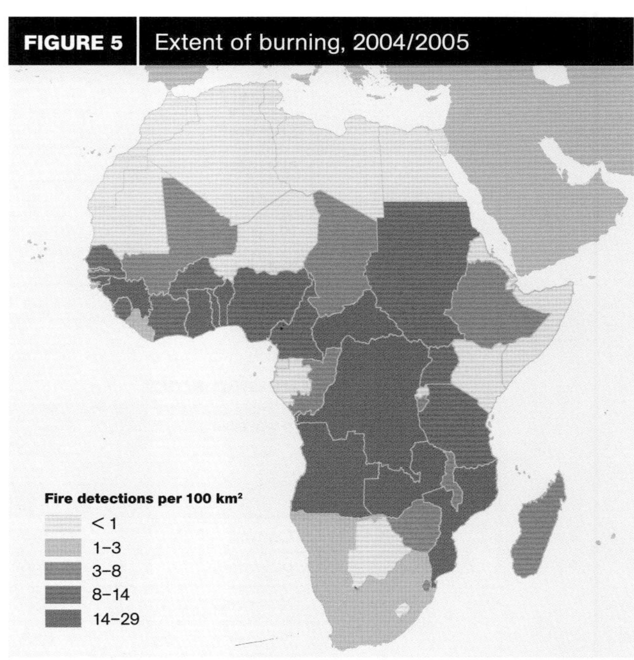


**NOTE:** Data derived from the Moderate Resolution Imaging Spectroradiometer (MODIS) satellite sensor at 1 km² resolution.
**SOURCE:** FIRMS (Fire Information for Resource Management System), University of Maryland, United States of America/United States National Aeronautics and Space Administration.

FRA 2005 by only five countries, and hence it was not possible to provide a regional overview of the situation and general trends. FAO has proposed a systematic process for improving data collection for FRA 2010 and has prepared forest pest profiles for Ghana, Kenya, Mauritius, Morocco, South Africa and the Sudan.

Despite the lack of data, there is no doubt that increasing problems with invasive insects, diseases and woody species have affected the productivity and vitality of African forests. The Forest Invasive Species Network for Africa was created to focus on these disturbances, with the mandate to coordinate collation and dissemination of information on forest invasive species in sub-Saharan Africa (www.fao.org/forestry/site/26951/en).

Accidental introductions of forest pests have affected industrial plantations of cypress and pine in East and Southern Africa for several decades. The cypress aphid, *Cinara cupressivora*, which affects Mexican cypress (*Cupressus lusitanica*) and pencil cedar (*Juniperus procera*), was first recorded in Malawi in 1986 and soon spread to neighbouring countries. It was estimated that the aphid had killed trees worth US$44 million as of 1990 and was causing the loss of a further US$14.6 million per year through reduction in annual growth increment (Murphy, 1996). Similarly, the European woodwasp (*Sirex noctilio*), accidentally introduced into South Africa, has infested pines and caused considerable impact on the industry. Special efforts are being made to prevent its further spread in the midlands of South Africa and to neighbouring countries.

New insect pests introduced into Africa within the past five years include *Coniothyrium zuluense* in Ethiopia, *Thaumastocoris australicus* and *Coryphodema tristis* in South Africa, *Leptocybe invasa* in Kenya, Uganda and the United Republic of Tanzania and *Cinara pinivora* in Malawi.

## PRODUCTIVE FUNCTIONS OF FOREST RESOURCES

The production of wood and non-wood forest products (NWFPs) is a very important function of African forests and woodlands and has great impact on socio-economic development. Some 30 percent of total forest area is designated primarily for production, compared with a global average of 34 percent.

The extent of forests designated for production is declining in Africa (Table 4). However, it is not clear if this should be considered a positive or negative trend in terms of sustainable forest management. It may be a sign that more area is excluded from productive purposes in order to enhance the conservation of biodiversity and other functions of forests; but it may also be an indication that productive forests are being cleared to convert land to non-forest uses. This is an area needing improved data in the context of FRA 2010.

TABLE 4
**Area of forest designated primarily for production**

| Subregion | Area (1 000 ha) | | | Annual change (1 000 ha) | |
|---|---|---|---|---|---|
| | 1990 | 2000 | 2005 | 1990–2000 | 2000–2005 |
| Central Africa | 45 268 | 43 790 | 41 992 | −148 | −360 |
| East Africa | 30 678 | 27 646 | 26 119 | −303 | −305 |
| Northern Africa | 35 067 | 32 899 | 31 331 | −217 | −313 |
| Southern Africa | 9 527 | 11 031 | 12 083 | 150 | 210 |
| West Africa | 27 789 | 24 548 | 23 134 | −324 | −283 |
| **Total Africa** | **148 329** | **139 913** | **134 658** | **−842** | **−1 051** |
| **World** | **1 324 549** | **1 281 612** | **1 256 266** | **−4 294** | **−5 069** |

TABLE 5
**Growing stock**

| Subregion | Growing stock | | | | | |
|---|---|---|---|---|---|---|
| | (million m³) | | | (m³/ha) | | |
| | 1990 | 2000 | 2005 | 1990 | 2000 | 2005 |
| Central Africa | 47 795 | 46 247 | 45 790 | 192 | 193 | 194 |
| East Africa | 4 989 | 4 616 | 4 446 | 56 | 57 | 58 |
| Northern Africa | 1 436 | 1 409 | 1 390 | 17 | 18 | 18 |
| Southern Africa | 6 669 | 6 292 | 6 102 | 35 | 36 | 36 |
| West Africa | 7 871 | 7 085 | 6 753 | 89 | 90 | 91 |
| **Total Africa** | **69 373** | **66 171** | **64 957** | **99** | **101** | **102** |
| **World** | **445 252** | **439 000** | **434 219** | **109** | **110** | **110** |

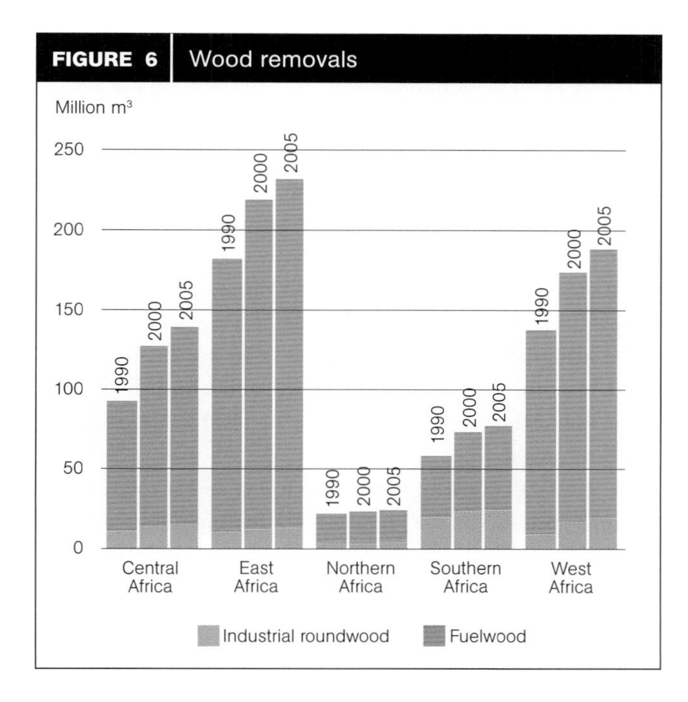


Growing stock is an important indicator of forest productivity. Although aggregated comparisons (growing stock per subregion or region) may not provide a clear picture, some general inferences can be made based on a global comparison. Country data suggest a significant decline in total growing stock in almost all countries (Table 5), although a few countries have registered an increase resulting from an increase in the area of forest plantations.

Another key issue for the productive functions of forests – given the declining trend in growing stock in most countries – is whether the level of wood removals exceeds the annual allowable cut. Almost 90 percent of the wood removals in Africa are used for fuel, compared with less than 40 percent in the world at large (Figure 6). For Africa as a whole, wood removals in 2005 were about 1 percent of growing stock. However there is considerable variation among regions, largely resulting from disparities in access to forest resources and the proportion of commercial species. For example, in West and Central Africa, the removal rate is about 0.06 percent of the estimated growing stock, while in Northern Africa it is over 7 percent.

In the absence of information on annual allowable harvests, it is difficult to conclude whether current removals are sustainable. The dominant use of wood in Africa is for fuel, and a large part of the demand is met from other wooded land and trees outside forests. Since market demand and forest access are key determinants of the intensity of wood removal, areas that are easily accessible are more intensively logged than remote ones.

## PROTECTIVE FUNCTIONS OF FOREST RESOURCES

For the 43 countries reporting, the extent of forest designated primarily for protection is about 4.5 percent of forest area and declined from 21.4 million hectares in 1990 to 20.6 million hectares in 2005, in line with the overall reduction in forest cover (Table 6).

However, not all countries use this designation, and some protective functions may be included under "multiple purpose" (Figure 7).

In proportion to the total area of forests, the reported extent of forests designated primarily for protection is low in Central Africa, but this is at least partly because of the relatively low level of reporting.

A number of countries have stepped up afforestation efforts with the primary objective of environmental protection. This includes afforestation of degraded areas for soil conservation, establishment of windbreaks and shelterbelts to protect agriculture areas, stabilization of sand dunes and urban and peri-urban planting to improve amenity values. In the 46 countries reporting on this activity, there was an increase in the extent of protective forest plantations of nearly 400 000 hectares (ha)

TABLE 6
**Area of forest designated primarily for protection**

| Subregion | Area (1 000 ha) | | | Annual change (1 000 ha) | |
|---|---|---|---|---|---|
| | 1990 | 2000 | 2005 | 1990–2000 | 2000–2005 |
| Central Africa | 368 | 746 | 651 | 38 | −19 |
| East Africa | 3 748 | 3 633 | 3 574 | −12 | −12 |
| Northern Africa | 3 645 | 3 819 | 3 861 | 17 | 8 |
| Southern Africa | 2 692 | 2 480 | 2 279 | −21 | −40 |
| West Africa | 10 939 | 10 610 | 10 247 | −33 | −72 |
| **Total Africa** | **21 392** | **21 287** | **20 613** | **−10** | **−135** |
| **World** | **296 598** | **335 541** | **347 217** | **3 894** | **2 335** |

**NOTE:** Fewer than 50 percent of the countries in Central Africa were able to report on this parameter.

FIGURE 7 Designated primary functions of forests, 2005

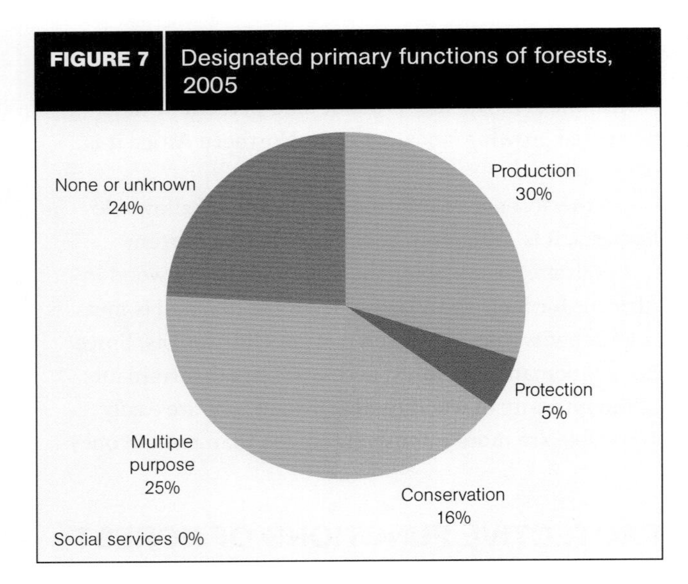


during 1990–2005. Most of the increase (over 87 percent) occurred in the poorly forested subregion of Northern Africa.

The total area of forest designated for protective functions shows a slight decrease for Africa as a whole, with Northern Africa being the only subregion with a slight increase. However, the area of protective forest plantations is increasing in four subregions and in the region as a whole. Overall, it is not possible to conclude that protective functions are improving; but in contrast with some of the other thematic elements, the trends are not alarmingly negative.

## SOCIO-ECONOMIC FUNCTIONS

The value of wood removals (fuelwood and industrial roundwood) in Africa increased from US$2.1 billion in 1990 to about US$3.9 billion in 2005. However, in spite of the relatively rapid growth in value, in 2005 its share in the global value of wood removals was only about 6.5 percent, while Africa accounted for about 16 percent of the world's forests.

Africa's share in the global value of industrial roundwood removals accounts for only about 4.7 percent, whereas its share in the value of fuelwood removal is about 22 percent. In fact, fuelwood value represented almost 35 percent of the total value of wood removals in 2005, although this proportion has been declining since 1990. No other region has recorded such a high share for the value of fuelwood removals in the total value of all wood removed.

The overall contribution of the forest sector to gross domestic product (GDP) registered a marginal increase from about US$7.3 billion in 1990 to about US$7.7 billion in 2000 (Figure 8). However, the forest sector share of total GDP has been declining over time, from about 1.7 percent in 1990 to about 1.5 percent in 2000. This decline is largely a result of the faster growth of other sectors, increasing GDP, while forestry's value added has not increased significantly. Within the forest sector, value added in the wood-processing and pulp-and-paper subsectors has remained more or less the same, and roundwood production (which includes industrial roundwood and fuelwood) accounts for almost 57 percent of value added.

This is in contrast to the global situation, where wood processing and pulp and paper together account for almost 78 percent of the value added, while the share of roundwood production is only about 22 percent. Africa's share of the forest-sector value added is about 2.2 percent of the global total. Its shares of value added in the wood-industries and pulp-and-paper subsectors are about 1.3 and

FIGURE 8 Trends in value added in the forest sector, 1990–2000

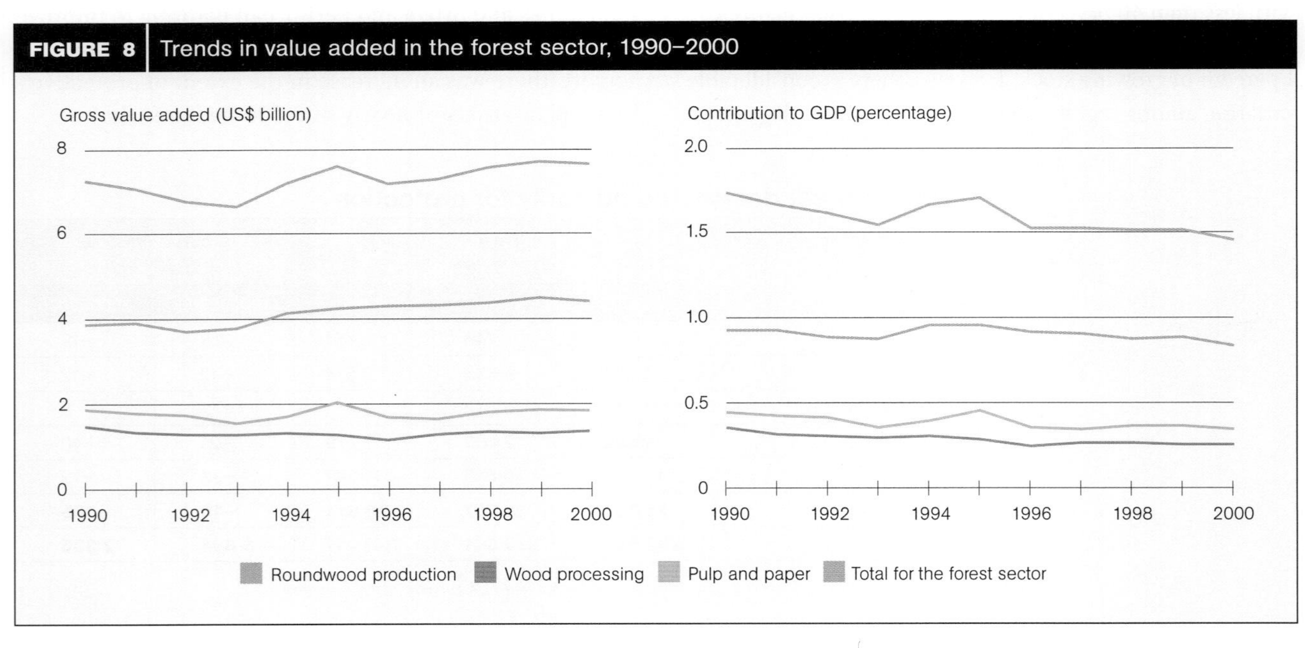

1.1 percent, respectively, while roundwood production is about 5.7 percent of the global value.

There are many countries in which the development of competitive wood-processing and pulp-and-paper subsectors has led to a high share in value added from the forest sector, although the value added generated in wood production is extremely low. This offers an important lesson – possession of a large tract of forests and increased wood production are neither necessary nor sufficient conditions for the existence of a vibrant forest industry. Increasingly, natural advantage (for example, the existence of vast tracts of forests) is being replaced by competitive advantage (Figures 9 and 10).

Employment in the formal forest sector in Africa increased from about 520 000 persons in 1990 to about 550 000 in 2000 (Figure 11). This increase is largely attributable to growth in employment in wood processing, which on average accounted for 60 percent of employment. However, following a significant increase from 1992 to 1995, there was an apparent decline in the late 1990s.

While there has been some growth in employment in the formal forest sector, the share of forestry in total employment has declined marginally from about 0.20 percent in 1990 to 0.16 percent in 2000.

A problem in assessing the socio-economic significance of the forest sector in Africa is the scarcity of data on production and employment in the informal sector. Microlevel studies suggest that the informal sector is predominant, but national statistics on income and employment emphasize the formal sector. Significant shares of wood production (particularly fuelwood) and processing (for example, pit-sawing, charcoal production, collection and trade of NWFPs) take place in the informal sector, and thus no national statistics are available. In some countries in the region, in particular in West and Central Africa, bushmeat is the most important single source of protein in the diet; yet this important NWFP is not usually reflected in official statistics.

The importance of the informal sector also raises some significant issues for progress towards sustainable forest management. Since those who operate in the informal sector often have no rights over the land and forests, most collection of wood and other products is "illegal" in the existing legal framework of most countries. In the context of ill-defined rights, there is little incentive to manage resources sustainably. Further, most of those dependent on the informal sector are poor, without the necessary resources to practise such

**FIGURE 9** Trends in net trade of forest products by subsector

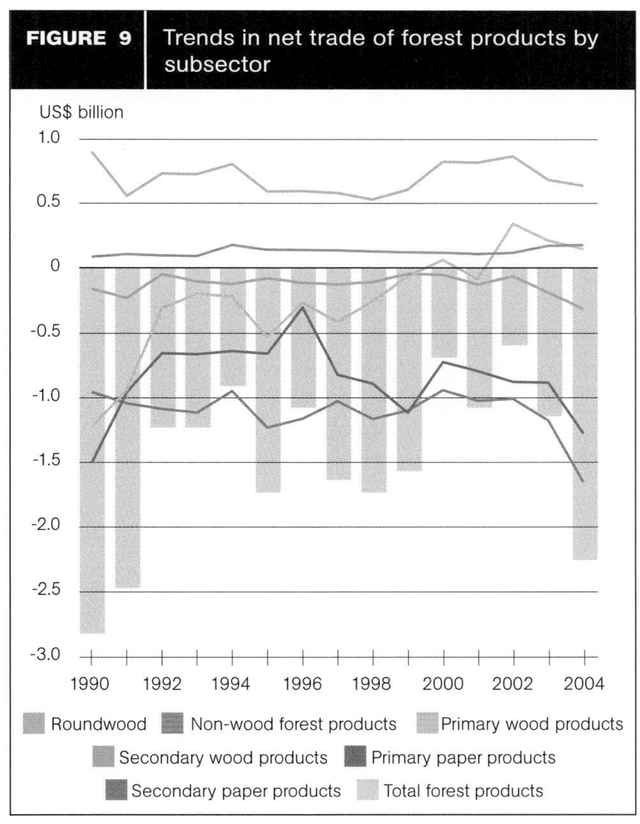


**NOTES:** A positive value indicates net export. A negative value indicates net import. Primary wood products include roundwood, sawnwood, wood-based panels and wood chips. Secondary wood products include wooden furniture, builders' joinery and carpentry. Primary paper products include pulp, paper and paperboard. Secondary paper products include packaging cartons, boxes and printed articles, including books and newspapers.

**SOURCES:** FAO, 2006b; United Nations, 2006.

**FIGURE 10** Trends in net trade of forest products by subregion

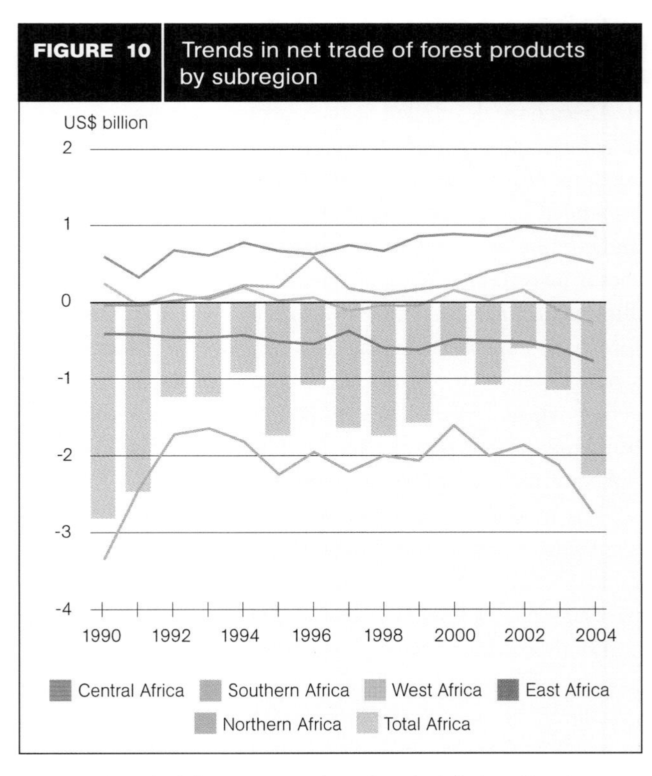


**NOTE:** A positive value indicates net export. A negative value indicates net import.

FIGURE 11 Employment in the formal forest sector

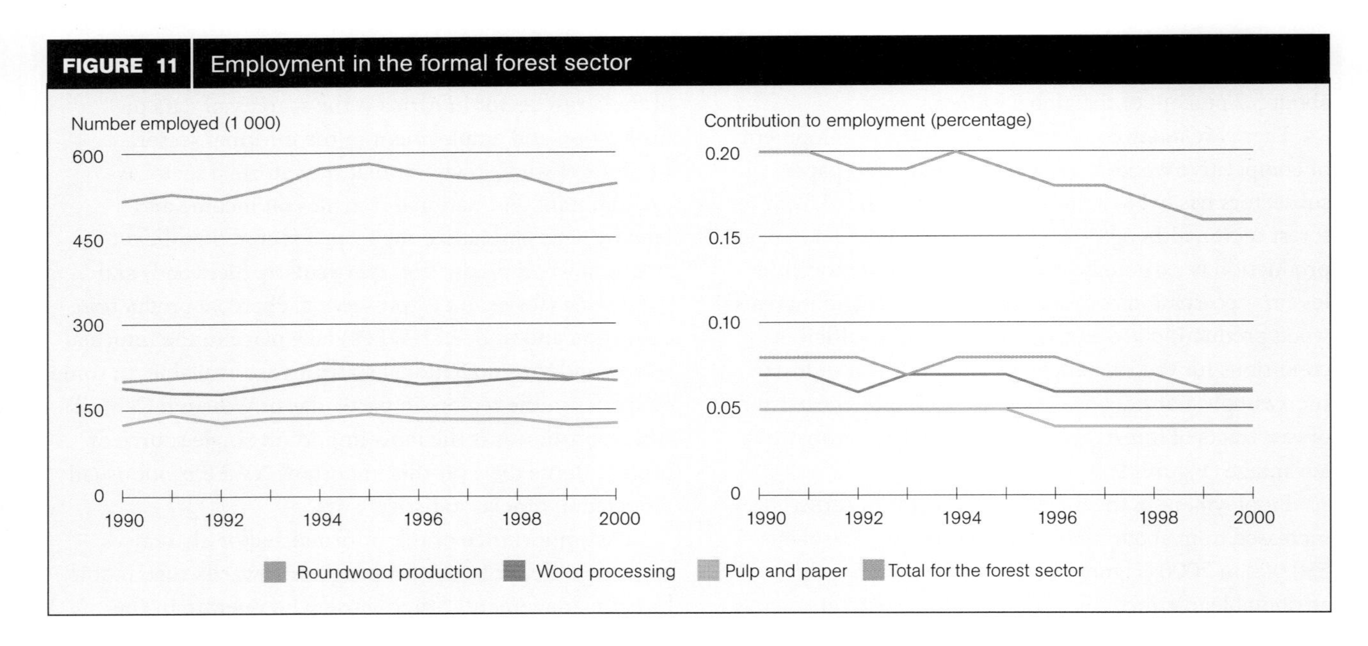


management. This would suggest that improvements in the functioning of the informal sector are needed in order to make progress towards sustainable forest management.

## LEGAL, POLICY AND INSTITUTIONAL FRAMEWORK

During the past decade, more than half the countries in Africa have developed or have been developing a new forest policy. The general trend is towards more sustainable, decentralized forest management, including enhanced access and management rights for local people and communities and the strengthening of private-sector investment. Several countries have made poverty alleviation a focus of their forest policy. In a few cases, forestry is considered in the national poverty reduction strategy. About two-thirds of the countries also have an active national forest programme in various stages of implementation, and 21 have established partnerships with the National Forest Programme Facility.

The implementation of new policies has been affected by obstacles: inadequate political support to the forest sector; weak capacity to implement processes that are participatory and involve cross-sectoral issues; and limited ability to mobilize external and internal financial resources to support key strategic actions.

Significant reforms of forest laws have taken place in many African countries. Since 1992, more than half the countries have developed new forestry laws or codes. While these vary in terms of their approaches and the depth to which they address particular issues, in general they evidence a broad trend towards:

- strengthening forest management planning;
- promoting sustainable forest management;
- strengthening the potential for community and private-sector forest management, including decentralization of responsibilities with greater local involvement; and
- recognition of environmental and biodiversity concerns, including forest protection.

Despite progress in improving forestry legislation, implementation and law enforcement remain weak in most countries. New legal provisions have been enacted in countries with political instability and weak political will, fragile civil society organizations, lack of administrative capacity and unfavourable local and national economies.

While new forest laws in some countries include provisions to transfer utilization and management rights to private individuals or communities, public administrations still play a dominant role in virtually every country in Africa. In many countries, national forestry agencies lack the means and capacity to fulfil the duties assigned by law.

At least one-third of the countries in Africa have undergone structural reforms of their forestry administrations in the past decade. In some countries, forestry departments were transferred to newly created environment ministries. There have been institutional reforms that included decentralization of management authority for forest resources and/or devolution of management rights to local people and the private sector. Implementation remains weak in many countries, and high rates of mortality from HIV/AIDS and other diseases have adversely affected institutions.

A number of countries have restructured their national agricultural research systems, including forest research. The prevailing trend is regionalization of agricultural research within countries, with regional programmes undertaken by multidisciplinary research teams. While justified and positive in many aspects in principle, this restructuring may, in fact, weaken national capacity in forest research by spreading limited national expertise too thinly. In some countries, coordination of such research at national and regional levels is not satisfactory. Government and donor funding of forest research has been declining over the past decade.

Forestry educational institutions in Africa vary widely in terms of funding support, number of graduates and quality of curricula. According to a survey in sub-Saharan countries (FAO, 2005a), forestry educational institutions from nine countries indicated that funding was intermittent, declining and came largely from national resources. Graduation at the forestry certificate level has dropped drastically, mainly because of low enrolment and the closure of certificate programmes. In general, forestry education needs are not properly identified, and plans are poorly articulated. Forestry authorities, the private sector and educational institutions need to engage in multipartner dialogue to improve forestry education planning. Several networks for forestry education have emerged, such as the African Network on Agroforestry Education (ANAFE) and a network of forestry and environmental education institutes in the Central African subregion, Réseau des institutions de formation forestière et environnementale d'Afrique centrale (RIFEAC).

A particularly positive development is the growth of regional cooperation at the policy level to address forest issues – through initiatives such as the Southern African Development Community (SADC), the Conference of Ministers in Charge of Forests in Central Africa (COMIFAC) and the New Partnership for Africa's Development (NEPAD).

Leaders in many African countries have demonstrated political commitment to support sustainable forest management through forestry laws, policies and national forest programmes. Weak capacity and inadequate resources continue to hamper efforts to implement these reforms effectively in many countries. Nonetheless, in this thematic element it can be concluded that significant progress is being made to establish a framework for sustainable forest management on which the other thematic elements can build.

## SUMMARY OF PROGRESS TOWARDS SUSTAINABLE FOREST MANAGEMENT

Progress towards sustainable forest management in Africa is slow and uneven. The legal and policy environment is improving in many countries, as evidenced by political commitment at the highest levels, by the development of national forest programmes throughout the region, and by progressive new forestry legislation in many countries. Regional partnerships such as NEPAD and COMIFAC provide a solid framework for action. However, the investment in forestry remains far below what is required, and the capacity to enforce laws and to implement programmes effectively remains weak in most countries. Some key concerns are summarized:

- Although the rate of forest cover loss is slowing slightly, on the whole the rate remains high. The extent of other wooded land is also declining.
- Afforestation and reforestation efforts fall short of compensating for the loss of natural forests. Most of these efforts are in countries with low forest cover (especially in Northern Africa).
- The area of primary forests in Africa is declining, but there has been some increase in the extent of area designated primarily for the conservation of biological diversity.
- The lack of reliable and consistent data over a sufficiently long period prevents any meaningful conclusion on the state of forest health and vitality.
- The total area designated primarily for protection has declined over the years, even though the percentage of protected forest has increased in some countries. There has been an increase in the extent of protective forest plantations, although much of this, again, is in Northern Africa, and consists primarily of countries with low forest cover.
- The value of wood removals has increased, but fuelwood still accounts for a larger share than in other regions. Official reports do not reflect actual removals, in view of the predominance of the informal sector. It is not likely that sustainable forest management will be achieved without taking action to address many of the issues contributing to a strong informal sector, including poverty and land tenure.
- Because the informal sector is absent from national economic statistics, the importance of forestry in the region is strongly understated in many official studies. In particular, the forest sector should be a key component of national efforts to reduce poverty.
- Perhaps the most positive trend is that most African countries have made legal, policy and institutional changes. However, the ability of institutions to implement sustainable forest management is limited, owing largely to the overall unfavourable social and economic situation.
- Information and communication on forest-sector issues remain weak and will require new approaches at the national level to open communication, reliable monitoring systems and sharing of information and experiences.

Overall, progress towards sustainable forest management in Africa is uneven. In comparison with most other regions, Africa lags behind. The challenge is to build on the positive trends and to take effective action to halt the most serious negative ones.

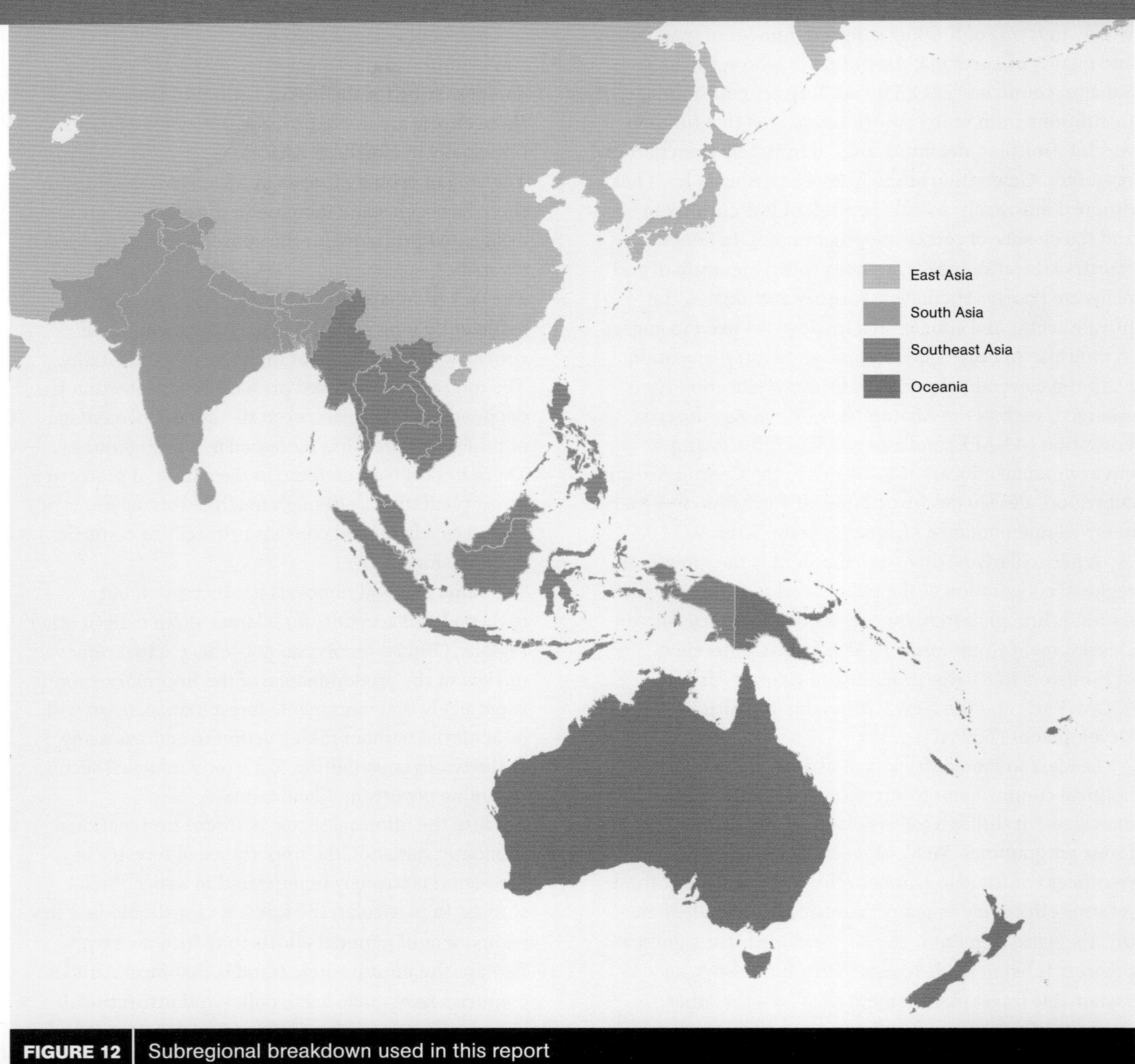


**FIGURE 12** Subregional breakdown used in this report

**East Asia:** China, Democratic People's Republic of Korea, Japan, Mongolia, Republic of Korea

**South Asia:** Bangladesh, Bhutan, India, Maldives, Nepal, Pakistan, Sri Lanka

**Southeast Asia:** Brunei, Cambodia, Indonesia, Lao People's Democratic Republic, Malaysia, Myanmar, Philippines, Singapore, Thailand, Timor-Leste, Viet Nam

**Oceania:** American Samoa, Australia, Cook Islands, Fiji, French Polynesia, Guam, Kiribati, Marshall Islands, Micronesia, Nauru, New Caledonia, New Zealand, Niue, Northern Mariana Islands, Palau, Papua New Guinea, Pitcairn, Samoa, Solomon Islands, Tokelau, Tonga, Tuvalu, Vanuatu, Wallis and Futuna Islands

**NOTE:** Statistics for the Russian Federation are included in the European region, where the capital is located. However, much of the forest area of the Russian Federation is technically in Asia, and we acknowledge that forest statistics for the region would increase significantly if the Russian Federation were included.

# Asia and the Pacific

## EXTENT OF FOREST RESOURCES

Forests and other wooded land together cover about one-third of the Asia and the Pacific region (Figure 13). Excluding the Russian Federation, forest area in 2005 was estimated at 734 million hectares, accounting for about 19 percent of global forest area. The region as a whole experienced a net increase in forest area of about 633 000 ha annually during 2000–2005 (Table 7). This is an important breakthrough, since the region had experienced a net loss of forest cover during the 1990s. The improvement was largely the result of an increase of more than 4 million hectares per year in China, which has been investing heavily in afforestation in recent years.

Bhutan, India and Viet Nam also increased their forest area from 2000 to 2005. However, most other countries experienced a net loss. Southeast Asia experienced the largest decline in forest area, with an annual net loss of forests of more than 2.8 million hectares per year, about the same rate as had occurred during the 1990s. The greatest forest loss occurred in Indonesia, almost 1.9 million hectares per year, followed by Myanmar, Cambodia, the Philippines, Malaysia and the Democratic People's Republic of Korea.

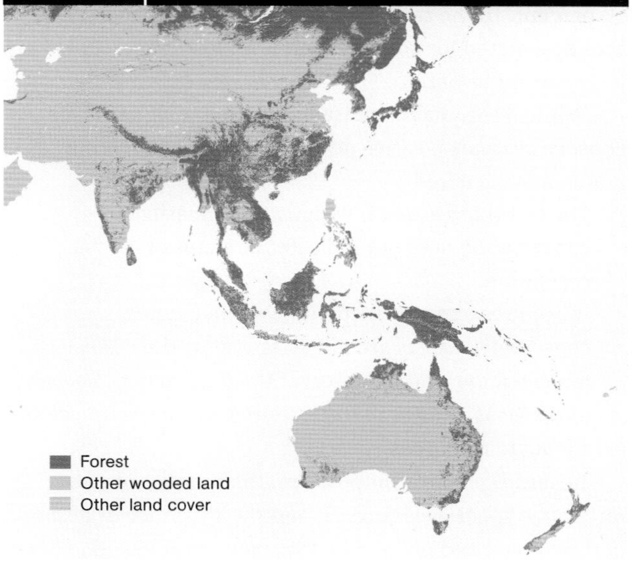

During the first five years of the twenty-first century, the variation among Asian countries in the net rate of change of forest area was dramatic; this variation is much more pronounced in Asia and the Pacific than in other regions. Several countries are losing forests at rates exceeding 1.5 percent per year; these are among the highest rates of loss in the world (Figure 14). On the other hand, forest area is increasing considerably in several countries, especially in China and Viet Nam.

Other wooded land is extensive, accounting for 13 percent of the land area in the region. The overall trend in other wooded land is downward, both in Asia and the Pacific and in the world, despite an increase in Southeast Asia. However, reporting on this category is not fully consistent from one country to the next, and it is difficult to monitor this category with remote sensing, so significant conclusions have not been drawn from the data.

**FIGURE 13** Extent of forest resources

**SOURCE:** FAO, 2001a.

TABLE 7
**Extent and change of forest area**

| Subregion | Area (1 000 ha) | | | Annual change (1 000 ha) | | Annual change rate (%) | |
|---|---|---|---|---|---|---|---|
| | 1990 | 2000 | 2005 | 1990–2000 | 2000–2005 | 1990–2000 | 2000–2005 |
| East Asia | 208 155 | 225 663 | 244 862 | 1 751 | 3 840 | 0.81 | 1.65 |
| South Asia | 77 551 | 79 678 | 79 239 | 213 | –88 | 0.27 | –0.11 |
| Southeast Asia | 245 605 | 217 702 | 203 887 | –2 790 | –2 763 | –1.20 | –1.30 |
| Oceania | 212 514 | 208 034 | 206 254 | –448 | –356 | –0.21 | –0.17 |
| **Total Asia and the Pacific** | **743 825** | **731 077** | **734 243** | **–1 275** | **633** | **–0.17** | **0.09** |
| **World** | **4 077 291** | **3 988 610** | **3 952 025** | **–8 868** | **–7 317** | **–0.22** | **–0.18** |

TABLE 8
**Area of forest plantations**

| Subregion | Area (1 000 ha) | | | Annual change (1 000 ha) | |
|---|---|---|---|---|---|
| | 1990 | 2000 | 2005 | 1990–2000 | 2000–2005 |
| East Asia | 29 531 | 35 518 | 43 166 | 599 | 1 530 |
| South Asia | 2 719 | 3 651 | 4 073 | 93 | 84 |
| Southeast Asia | 10 046 | 11 550 | 12 561 | 150 | 202 |
| Oceania | 2 447 | 3 459 | 3 833 | 101 | 75 |
| **Total Asia and the Pacific** | **44 743** | **54 178** | **63 633** | **943** | **1 891** |
| **World** | **101 234** | **125 525** | **139 466** | **2 424** | **2 788** |

**FIGURE 14** Forest change rates by country or area, 2000–2005

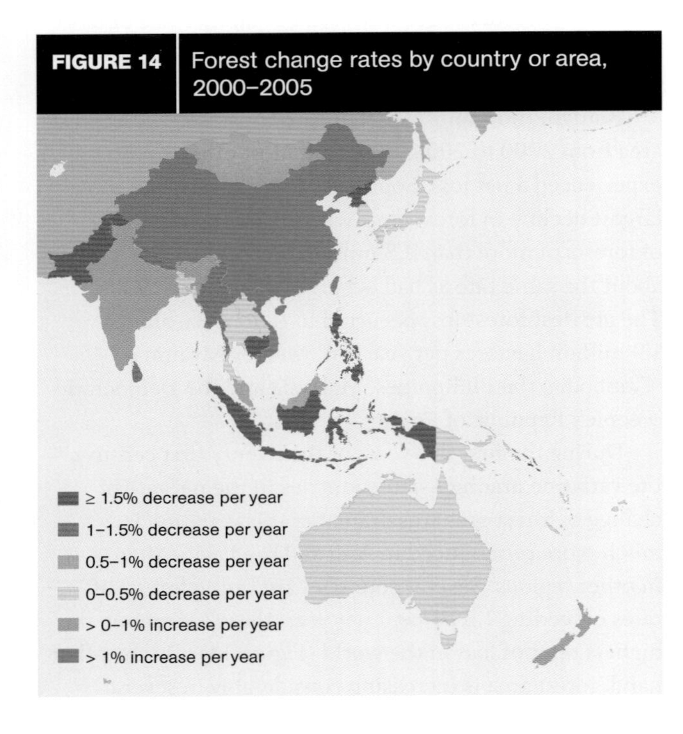


All subregions within the broader Asia and the Pacific region experienced a substantial increase in forest plantations during 2000–2005, continuing the trend from the 1990s (Table 8). China led the way; Viet Nam, India, Indonesia, Australia, the Republic of Korea, Myanmar, the Lao People's Democratic Republic and New Zealand have also made significant investments in forest plantations in recent years.

Net forest cover in Asia and the Pacific is increasing, which is an encouraging sign. This net increase at the regional level is built mainly on large investments in forest plantations in several countries. However, the growth in plantations does not negate the continued loss of natural forests.

## BIOLOGICAL DIVERSITY

In East Asia, the area of primary forests is fairly stable, having declined slightly in the 1990s and increased slightly since 2000. In South Asia, a negative trend has continued and accelerated over the past 15 years. In Southeast Asia, the negative trend is consistent and very disturbing, exceeding a 2 percent loss of primary forest per year. In Oceania, a recovery of primary forests in the 1990s has been replaced by a negative trend since 2000.

The area of forest designated primarily for conservation has been increasing in the Asia and the Pacific region as a whole since 1990 (Table 9). Only in Oceania has there been a slight decrease since 2000, but there have been significant increases in Southeast Asia and East Asia. For the region as a whole, the area of forest designated for conservation of biological diversity is slightly over 10 percent. In the tropical forests of Southeast Asia, the area designated for conservation is almost 20 percent. This is a heartening trend.

While a regional increase in the area designated for conservation is a positive development, the following points may be noted:

- The fact that an area is designated primarily for conservation does not indicate the status of its vegetation.
- In some instances, the policies for managing conservation areas are not clear and/or the management of the conservation areas may not be very effective as a result of institutional weakness or lack of adequate resources.

Regarding forest composition, the number of native forest tree species (Figure 15) and the existence or absence of threatened and endangered species, reliable and comprehensive data for these parameters are not available for most countries or for the region as a whole. About half the countries provided information on the composition and diversity of tree species.

Based on this limited information, it can be seen that forest composition and the distribution of species differ widely within the region. As might be expected, countries with moist tropical forests have more tree species than more temperate countries. For example, the Philippines has an estimated 3 000 native tree species, compared with 105 in Bhutan (or, for that matter, with 180 in Canada).

The extent to which a tree species faces the threat of extinction is another useful parameter for assessing forest

TABLE 9
**Area of forest designated primarily for conservation**

| Subregion | Area (1 000 ha) | | | Annual change (1 000 ha) | |
|---|---|---|---|---|---|
| | 1990 | 2000 | 2005 | 1990–2000 | 2000–2005 |
| East Asia | 10 338 | 10 847 | 11 479 | 51 | 126 |
| South Asia | 14 911 | 16 966 | 17 265 | 205 | 60 |
| Southeast Asia | 31 814 | 35 574 | 40 025 | 376 | 890 |
| Oceania | 6 709 | 7 968 | 7 948 | 126 | –4 |
| **Total Asia and the Pacific** | **63 772** | **71 355** | **76 717** | **758** | **1 072** |
| **World** | **298 424** | **361 092** | **394 283** | **6 267** | **6 638** |

**FIGURE 15** Number of native tree species

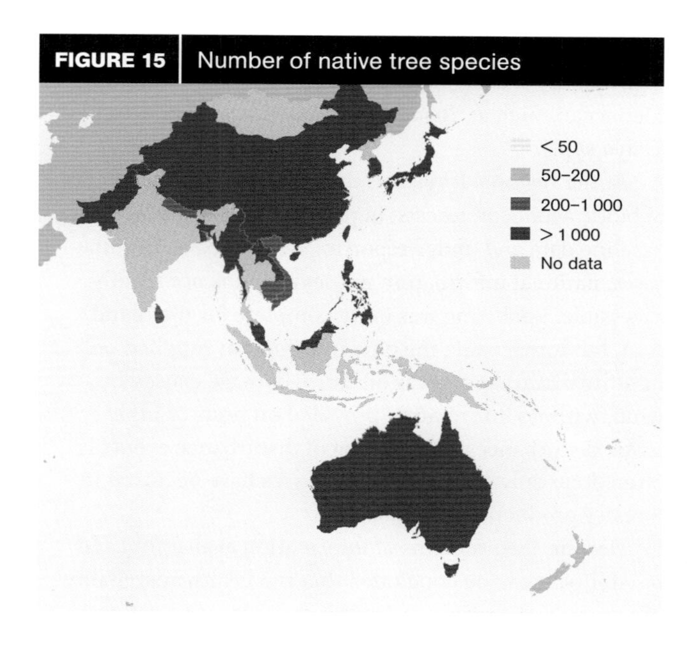


biological diversity. In Asia and the Pacific, Indonesia has the largest number of critically endangered tree species (IUCN, 2000, 2004), with 122 such species, followed by Sri Lanka and Japan. Malaysia has the largest number of vulnerable species – 403. Asia and the Pacific as a whole ranks as one of the regions with the largest number of endangered and vulnerable species.

## FOREST HEALTH AND VITALITY

Many of the countries did not report on forest fire (or the more inclusive "wildland fire"). Consequently, the estimated area burned in Australia dominates the regional statistics (Figure 16).

The following trends were observed in comparing two periods, 1988–1992 and 2000–2004 (FAO, 2006d):

- In East Asia, wildland fire increased in terms of scale, frequency, extent of damage and cost of fire suppression. Factors contributing to this trend include increases in periods of drought, climatic variability and population.
- In South Asia, fire is commonly used to clear land, and runaway agricultural fires are the cause of most uncontrolled wildland fires. Over 90 percent of the area burned in South Asia is in India, where significant efforts have been made to prevent and manage wildland fire over the past 20 years.
- In Southeast Asia, wildfires dominated the headlines in the late 1990s, when hot, dry climatic conditions favoured the outbreak of thousands of uncontrolled fires, which burned for months, resulting in smoke pollution that caused serious health and economic damage to the region. This led to the Association of Southeast Asian Nations Agreement on Transboundary Haze Pollution, signed by all association member countries in 2002, which entered into force in 2003. However, Indonesia, which has the most significant fire problem in Southeast Asia, has not ratified the agreement.
- In Oceania, the record fire season of 2002/03 in Australia was one of the largest disasters in the country's history, resulting in loss of human life and astronomical economic damage. Many of the fires were set by arsonists, and the combination of heat and drought were such that many fires were not brought under control for several weeks.

**FIGURE 16** Average annual area burned by wildland fire, 2000–2004

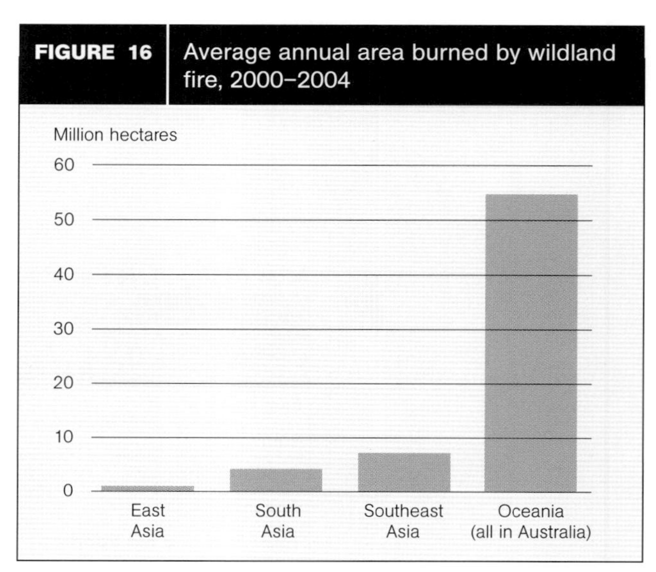


**SOURCE:** FAO, 2006d.

The problem of fire is increasing in all subregions. The problems in Southeast Asia are perhaps the most salient in that moist tropical forests were previously considered sheltered from fire. However, in the last two decades the subregion has experienced huge fires, mainly resulting from poor logging and agricultural practices. In the region as a whole, fires have caused massive problems, affecting human health and causing economic losses to the tourism and transport industries. Appropriate preventive measures are needed.

While fire gets the most attention in the media, studies indicate that forest pests and other disturbances may have a more widespread impact than fire in Asia and the Pacific. The Asia–Pacific Forest Invasive Species Network (APFISN) has been created to address these concerns.

Disturbances to forests by pests and abiotic factors significantly affect productivity. Preliminary research indicates that economic losses resulting from invasive plant species alone may total hundreds of billions of dollars.

More than 10 million hectares of forest were reported to be affected by insect pests annually (average 1998–2002), and more than 9 million hectares by diseases during the same period (Figure 17).

**FIGURE 17** Forest disturbances, 1998–2002

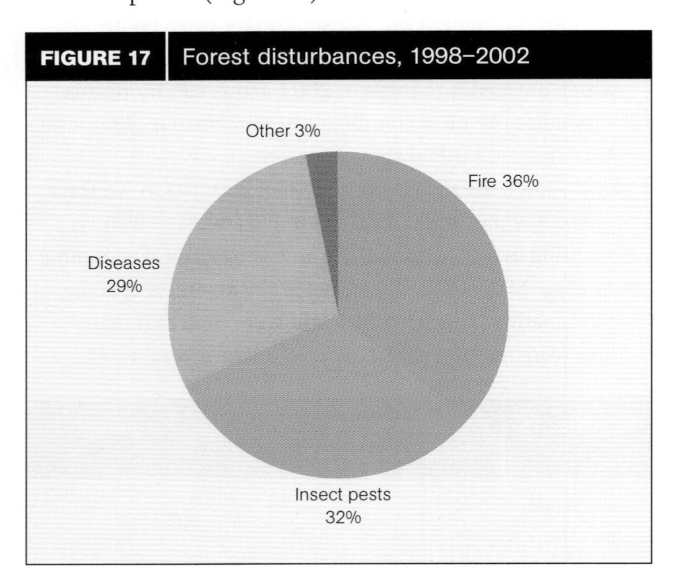


Serious outbreaks of *Anophlophora glabripennis* (Asian longhorn beetle) and *Dendrolimus sibiricus* (Siberian caterpillar) have caused significant concern both within the region and to international trade partners. Eucalyptus rust (*Puccinia psidii*), which is considered the most serious threat to *Eucalyptus* plantations worldwide, was the subject of an international workshop in Bangkok in October 2004. Other serious forest pests in the Asia and the Pacific region include *Heteropsylla cubana* (which damages some species of Fabaceae, including *Leucaena leucocephala*), *Corticium salmonicolor* (which damages a wide range of hosts, including *Acacia* spp. and eucalypts) and *Hypsipyla robusta* (which is a major pest of some high-quality timber species, particularly of the family Meliaceae, such as *Toona* spp., *Swietenia* spp. and *Khaya* spp.).

At the regional level, it is difficult to assess the effects of biotic agents on forests, in part because of the lack of baseline data and under-reporting of outbreaks. In some cases, national information may exist but is not readily accessible. Reporting was quite complete for mainland Asia, but for Oceania the only information supplied on non-fire disturbances was on storm damage caused by wind, with no information provided on pests or other biotic disturbances. Monitoring of disturbance events is often done only after significant losses have occurred in forestry production or trade.

Despite the problems of information availability and reliability, it can be concluded that the health and vitality of forests in the region are under stress from insect pests, disease, invasive plants and uncontrolled fire. One of the keys to sustainable forest management is to improve the understanding of these processes and the capability to manage and control them.

## PRODUCTIVE FUNCTIONS OF FOREST RESOURCES

In Asia and the Pacific, 37 percent of the total forest area is designated primarily for production, as compared with the global average of 34 percent (Table 10).

TABLE 10
**Area of forest designated primarily for production**

| Subregion | Area (1 000 ha) | | | Annual change (1 000 ha) | |
|---|---|---|---|---|---|
| | 1990 | 2000 | 2005 | 1990–2000 | 2000–2005 |
| East Asia | 126 821 | 119 688 | 125 488 | –713 | 1 160 |
| South Asia | 18 061 | 16 545 | 16 084 | –152 | –92 |
| Southeast Asia | 112 289 | 115 740 | 104 014 | 345 | –2 345 |
| Oceania | 5 651 | 9 371 | 9 261 | 372 | –22 |
| **Total Asia and the Pacific** | **262 822** | **261 344** | **254 848** | **–148** | **–1 299** |
| **World** | **1 324 549** | **1 281 612** | **1 256 266** | **–4 294** | **–5 069** |

At the regional level, the extent of forests designated for production was fairly stable in the 1990s, but has declined in the past five years. The downward trend has occurred both in Asia and the Pacific and in the world at large. However, it is difficult to ascertain whether this is a negative trend. It may be a sign that more area is excluded from productive purposes and set aside for conservation, or it may be an indication that productive forests are being cleared to convert land to non-forest uses.

Growing stock is another indicator of forest productivity (Table 11). Country data suggest a decline in total growing stock in many countries, with the exception of countries with large investments in forest plantations. The net result at the regional level is a modest decline in total growing stock in cubic metres and in cubic metres per hectare.

Regarding trends in wood removals (Figure 18), about 40 percent of the wood in the region is used for fuel, the same as the global average. However, the importance of wood for fuel is highly variable throughout the region: in South Asia, 89 percent of wood is used for fuel, about the same as in Africa; the figure drops to 64 percent in Southeast Asia, 33 percent in East Asia and 16 percent in Oceania.

For Asia and the Pacific as a whole, wood removals in 2005 were about 0.76 percent of growing stock, higher than the global average of 0.69 percent, but lower than Africa at 0.90 percent.

Within the region, the highest rates of wood removals as a percentage of growing stock are in East Asia and Oceania, 0.87 percent. The lowest rate is in Southeast Asia, 0.61 percent, whereas South Asia is at 0.76 percent, equal to the regional average. Variations among regions and subregions result from such factors as access, proportion of commercial species, effectiveness of management controls and the supply and demand for wood.

In consonance with global trends, productive forests in the Asia and the Pacific region have declined in the recent past. This trend is further reflected in terms of growing stock and wood removals, both industrial roundwood and fuelwood. However, in the absence of information on annual allowable harvests, it is difficult to establish if current removals are sustainable. Because market demand and forest access are key determinants of the intensity of removals, areas that are easily accessible are more intensively logged than remote ones.

**FIGURE 18** Wood removals

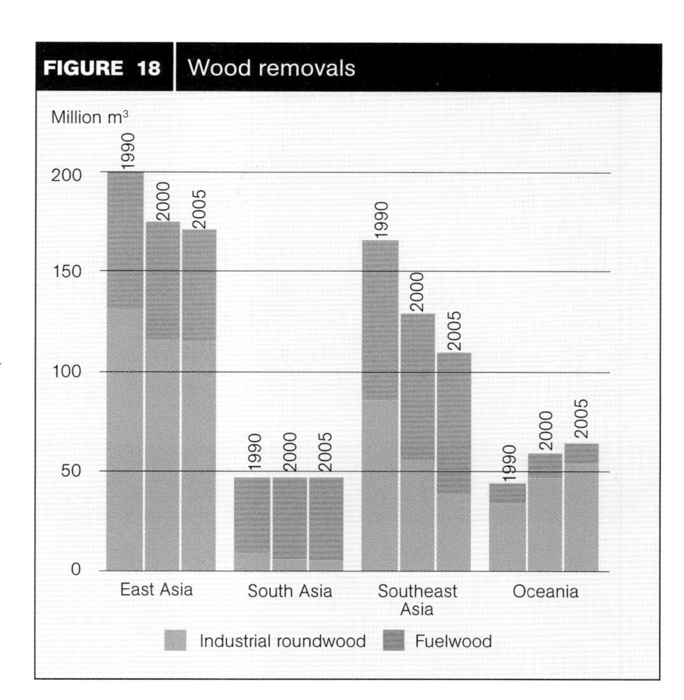


## PROTECTIVE FUNCTIONS OF FOREST RESOURCES

The area designated for protection has been increasing for the region as a whole, resulting mainly from increases exceeding 4 percent per year in East Asia (Table 12). However, most countries in Oceania did not report on this parameter and, in fact, not all countries use this designation. Thus some protective functions may be included under "multiple purpose" (Figure 19).

Various countries have stepped up afforestation efforts with the primary objective of environmental protection. This includes afforestation of degraded areas for soil conservation, establishment of windbreaks and shelterbelts to protect agriculture areas, stabilization of sand dunes, and urban and peri-urban planting to improve amenity

TABLE 11
**Growing stock**

| Subregion | Growing stock | | | | | |
|---|---|---|---|---|---|---|
| | *(million m³)* | | | *(m³/ha)* | | |
| | 1990 | 2000 | 2005 | 1990 | 2000 | 2005 |
| East Asia | 15 850 | 18 433 | 19 743 | 76 | 82 | 81 |
| South Asia | 5 714 | 6 237 | 6 223 | 74 | 78 | 79 |
| Southeast Asia | 26 909 | 21 063 | 17 981 | 110 | 97 | 88 |
| Oceania | 7 593 | 7 428 | 7 361 | 36 | 36 | 36 |
| **Total Asia and the Pacific** | **56 066** | **53 161** | **51 308** | **75** | **73** | **70** |
| **World** | **445 252** | **439 000** | **434 219** | **109** | **110** | **110** |

TABLE 12
**Area of forest designated primarily for protection**

| Subregion | Area (1 000 ha) | | | Annual change (1 000 ha) | |
|---|---|---|---|---|---|
| | 1990 | 2000 | 2005 | 1990–2000 | 2000–2005 |
| East Asia | 34 763 | 55 424 | 66 992 | 2 066 | 2 314 |
| South Asia | 12 065 | 12 021 | 11 991 | –4 | –6 |
| Southeast Asia | 45 357 | 46 886 | 47 106 | 153 | 44 |
| Oceania | 413 | 450 | 467 | 4 | 3 |
| **Total Asia and the Pacific** | **92 598** | **114 780** | **126 556** | **2 218** | **2 355** |
| **World** | **296 598** | **335 541** | **347 217** | **3 894** | **2 335** |

**FIGURE 19** Designated primary functions of forests, 2005

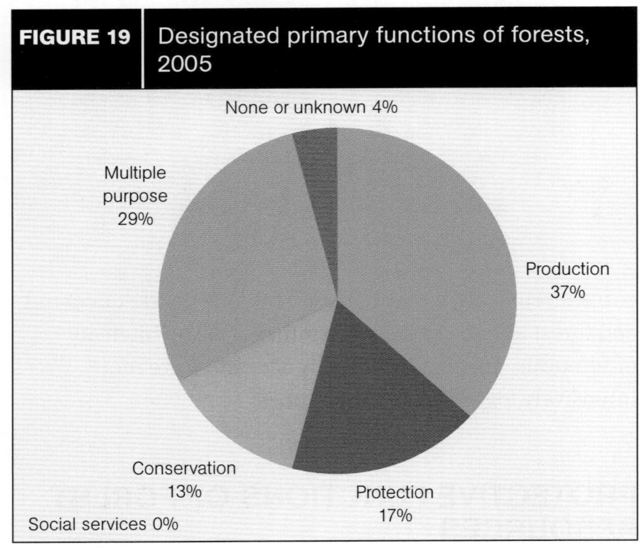


values. The overall increase in this parameter in recent years has been led by East Asia and South Asia.

A number of Asian countries are increasing the area of forest designated for protection and of forest plantations for protective purposes. However, the benefits of these protective functions have yet to be quantified or valued in financial terms and are rarely taken into account in assessing forest benefits. While it may be difficult, there is a need to develop markets for the protective functions of forests.

## SOCIO-ECONOMIC FUNCTIONS

There was a significant drop in the value of wood removals in Asia and the Pacific during the 1990s, primarily owing to the economic downturn late in the decade. The region accounts for about 24 percent of the forest sector's contribution to the global economy (roundwood production, wood-processing industries and pulp and paper). Including Oceania, the value added in Asia and the Pacific is about the same as that of Europe. Moreover, the contribution of the forest sector to total GDP is about the same as in the world at large, 1.2 percent.

The forest sector contributed an estimated US$85 billion to the economies of Asian countries in the year 2000 and more than US$5 billion to the economies of Oceania. During the 1990s, the value added in the forest sector of Asia and the Pacific remained relatively stable in real terms (Figure 20). In many countries, the value added in roundwood production is low, but the

**FIGURE 20** Trends in value added in the forest sector, 1990–2000

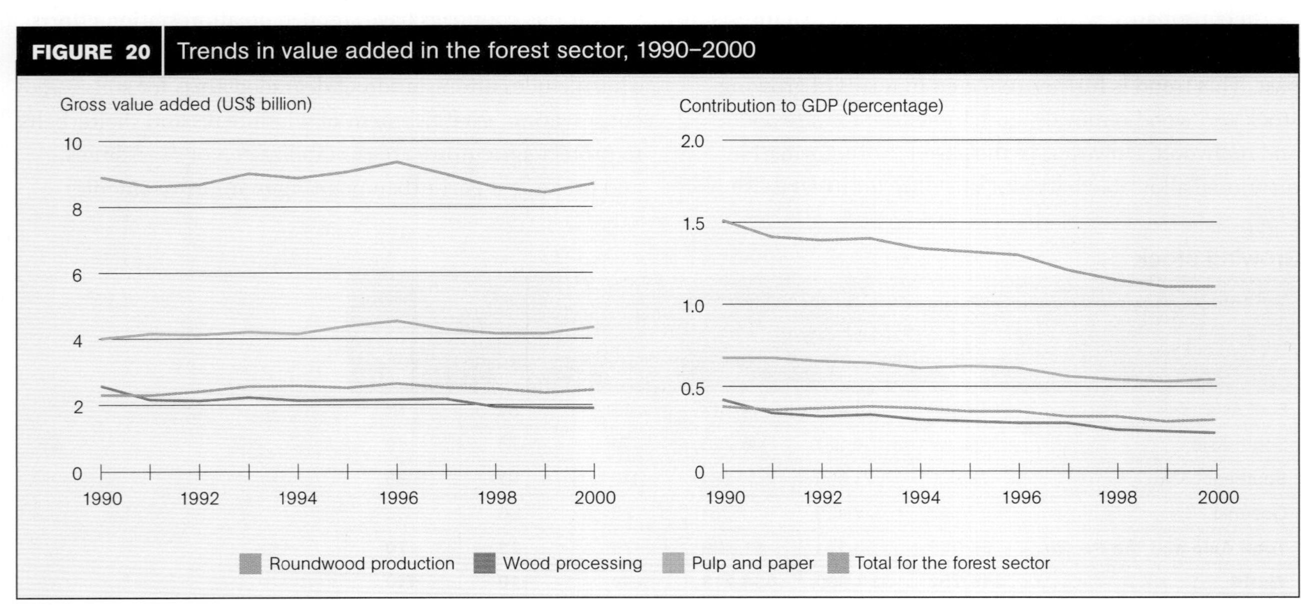

development of competitive wood-processing and pulp-and-paper subsectors has made an important contribution to economic growth.

Faster developments of other sectors of the economy in all subregions led to a decline in the forest sector's contribution to GDP. This trend is occurring in most regions of the world, with the exception of Latin America and the Caribbean.

The Asia and the Pacific region is the biggest net importer of forest products in the world. But the gap between imports and exports has remained relatively stable at about US$15 billion since the late 1990s (Figures 21 and 22). A rapidly developing secondary wood-processing sector (furniture, etc.), based on imported primary products and plantation timber, suggests that this trend will continue.

The region is the largest exporter of NWFPs (especially bamboo and rattan), amounting to some US$2–3 billion annually.

A problem in assessing the socio-economic significance of the forestry sector in Asia and the Pacific is the scarcity of data on production and employment in the informal sector. National statistics on income and employment emphasize the formal sector (Figure 23), while microlevel studies suggest that the informal sector is predominant.

**FIGURE 21** Trends in net trade of forest products by subsector

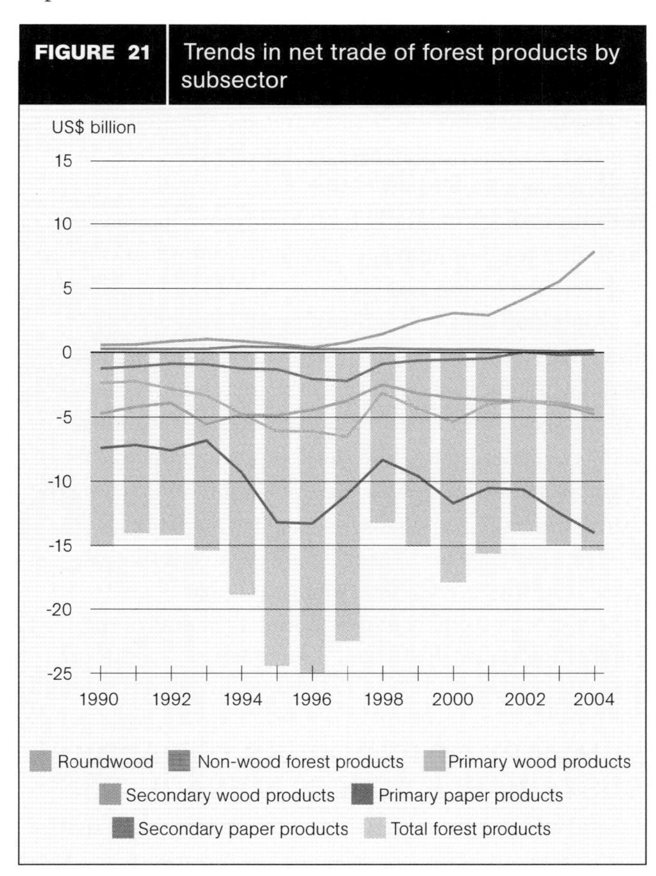


**NOTE:** A positive value indicates net export. A negative value indicates net import.

The importance of the informal sector also raises significant issues for progress towards sustainable forest management. Since those who operate in the informal sector often have no rights over the land and forests, their collection of wood and other forest products is often "illegal" in the existing legal framework of most countries. In the context of ill-defined rights, there is little incentive to manage the resources sustainably. Further, most of those dependent on the informal sector are poor, without the necessary resources to practise sustainable management. This would suggest that improvements in the informal sector are needed in order to make progress towards sustainable forest management.

Trends in forest-related socio-economic parameters suggest that the forest sector is likely to remain an important contributor to sustainable development. The combination of cheap labour, growing economies and consumer markets, and global trade possibilities creates a good basis for development.

## LEGAL, POLICY AND INSTITUTIONAL FRAMEWORK

Most countries in the region have a relatively sound legislative and policy foundation from which to implement sustainable forest management, and a majority have updated their forest policies in the past 15 years. Examples of policy changes since 2000 include policies to strengthen community involvement in Bhutan; new forest policies in Cambodia and Pakistan; the implementation of national forest programmes in India,

**FIGURE 22** Trends in net trade of forest products by subregion

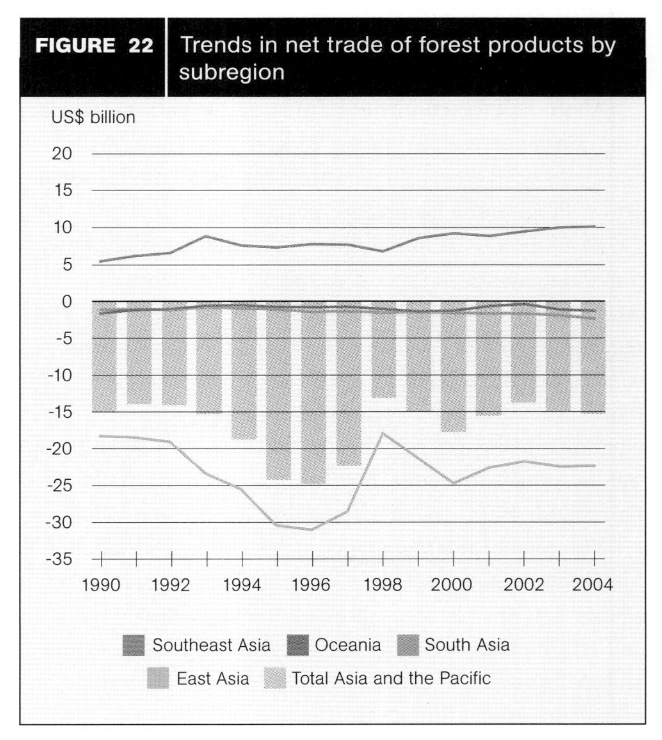


**NOTE:** A positive value indicates net export. A negative value indicates net import.

**FIGURE 23** Employment in the formal forest sector

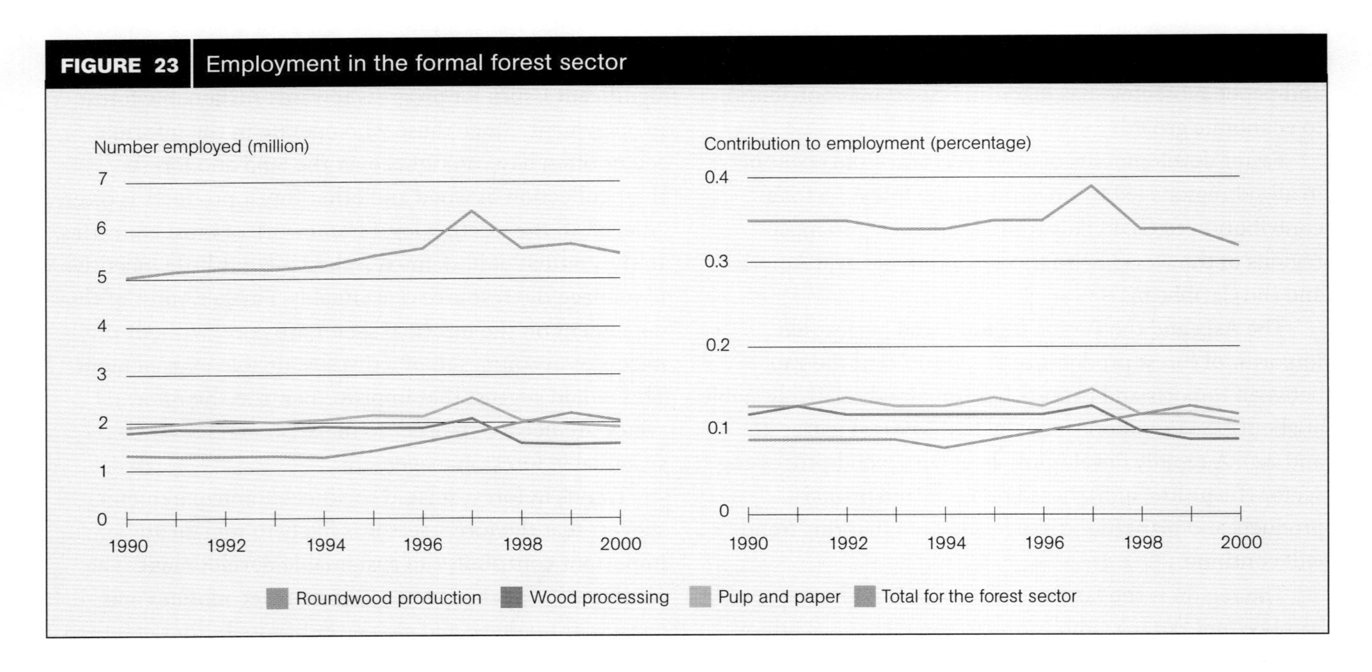


Indonesia, Mongolia and Nepal; the development of regional forest agreements in Australia; and a new National Forest Strategy (2006–2020) in Viet Nam.

A number of countries are moving towards policies that encompass participatory forestry and devolution and decentralization of forest management responsibilities. Some countries, such as Cambodia and Nepal, have focused on poverty reduction in their forest policy. However, despite the generally positive trend, in many countries policy objectives have not been achieved because of budget shortages, weak institutional capacity and governance problems. Some countries are making efforts to reinvent their forestry institutions (Box 2).

About half the countries have an active national forest programme, in various stages of implementation. Eight countries have established partnerships with the National Forest Programme Facility.

Legislation is the most important tool for translating policy statements into action. Most countries have a combination of policy statements, laws and programmes that regulate and orient the use of forests and the development of forest activities. Some countries have made efforts to modernize legislation to support economic, social and environmental policy frameworks (FAO, 2006e). Since 2000, significant new legislation has been enacted in Australia, Bangladesh, Bhutan, India, Mongolia, Vanuatu and Viet Nam.

Throughout the region, countries are devolving forest management responsibilities to local or provincial agencies, to the private sector and to community groups and NGOs. The private sector is increasing in importance, and many countries are trying to decentralize forest management and find more effective approaches to involving civil society.

At the regional level, significant progress has been made in strengthening institutions to support improved forest management. Regional institutions that have developed include the Asia–Pacific Association of Forestry Research Institutions, Asia–Pacific Agroforestry Network, APFISN and the Regional Community Forestry Training Center for Asia and the Pacific.

NGOs potentially play an important role in the forestry sector. NGO involvement in national forest programmes has increased from the 1990s onwards, and many have established networks to raise awareness, disseminate research and provide advice on forest conservation. In countries in which communities have a "hands-on" role in forest management, e.g. India and Nepal, institutional structures are developing to ensure a coordinated voice in macrolevel decision-making.

An important trend is the increased availability and accessibility of information, owing to the Internet and to the willingness of countries to share their forestry experiences. This has strengthened forest institutions in many countries. In the FRA 2005 reporting process, for example, countries in Asia and the Pacific were among the world leaders in responsiveness and participation.

The challenge for the region will be to ensure that some countries are not left behind, and that benefits are extended equitably to the poorest segments of the population, especially in rural forest areas.

## SUMMARY OF PROGRESS TOWARDS SUSTAINABLE FOREST MANAGEMENT

An encouraging trend in Asia and the Pacific is one that encompasses much more than just the forest sector – it is the high rate of economic growth in key countries in the

## BOX 2 Reinventing public forestry institutions

All over the world, forestry institutions are under pressure to adapt better to the environment in which they operate. Adapt and reinvent or fade into irrelevance is the norm in an increasingly competitive environment. While many countries have reformed forest policies and legislation, implementation lags behind because of institutional rigidity.

In the past, forest management in most countries was dominated by the public sector. This has changed in recent decades, as the private sector, local communities, farmers, etc. play an increasing role in all aspects of forestry. Institutional changes such as privatization, community forest management and an array of different partnerships reflect the range of options being pursued.

The fundamental driver of long-term change is evolution in the values, beliefs and perceptions of society. Drivers of institutional change include the following:

- **Macroeconomic policies** (often influenced by political ideologies). Economic liberalization and scaling down of government involvement, often to reduce budget deficits, have led to major changes in institutional arrangements in forestry. Social policies to reduce poverty and promote rural development have driven shifts towards greater involvement of local communities in forest management.
- **Changes in markets.** More flexible parastatal agencies (corporations, boards, enterprises) have been established to provide more flexibility for operating efficiently in a commercial environment.
- **Technological changes.** The increasing flow and volume of information make it possible to bypass lines of command and flatten organizational structures, and a more informed public is demanding efficiency, transparency and social and environmental responsibility.

Public-sector forestry agencies have often initiated structural changes themselves, largely out of resource constraints. Devolution of management responsibility to local levels is sometimes driven by the declining human and financial capacity of institutions and the need to reduce management costs.

The degree of change depends on the circumstances – from adaptation of functions and structure to external changes, to deep change involving revisiting the institution's core values and mission, followed by appropriate functional and structural changes.

Striking the right balance between stability and change is a major challenge facing forestry organizations. Change is necessary and inevitable, but some level of stability is also important, especially for consistency in the implementation of forest policies and, more importantly, to retain institutional capacity. Instability from too-frequent changes can promote staff attrition and undermine the accumulation of knowledge and experience as well as the development of institutional memory. People need to be an integral part of the change process.

Reinventing institutions is difficult and can be costly. Ideally, institutions should develop as learning organizations in tune with the needs of society. Addressing the human dimension of change is the most complex and least often successful aspect.

**Some examples from Asia and the Pacific**

An example of radical change is the privatization of forest plantations in New Zealand, largely triggered by economic liberalization policies. Less radical, but still substantial reforms – establishing more flexible autonomous parastatal agencies for specific activities, particularly in the realm of commercial forestry – have occurred in China, Fiji, India and Myanmar.

Divesting responsibilities to local communities is another major institutional development in Asia, particularly notable in several countries: joint forest management in India, management by forest users' groups in Nepal and community-based forest management in the Philippines.

There are also a number of instances in which government forestry agencies have brought production functions under more flexible autonomous institutions, overcoming constraints stemming from governmental rules and regulations. Research and development is another area that has been reinvented to respond to the special needs of scientific work (for example, the Forest Research Institute of Malaysia).

region, especially in the two largest countries, China and India. Many experts believe that this growth will have a positive impact on the economies of other countries. It is already having an impact on forestry in terms of demand for both forest products and service functions of forests.

- Primary forests continue to decline at a rapid rate in many countries, especially in Southeast Asia. Illegal logging continues in several countries, particularly in selected areas with high-value timber. Problems are most acute in countries that are not benefiting from economic growth, because such growth helps provide resources to strengthen institutions.
- While the net forest area in most countries of the region continues to decline, several countries are increasing their forest area as a result of investments in afforestation and rehabilitation.
- Economic development creates problems as well as opportunities. It is a challenge to ensure that commercial timber harvesting is done with care so that damage to the forest is minimized. Several Asian countries are implementing regional and national codes of forest harvesting practices to deal with this problem.
- High rates of forest plantation can result in a false sense of progress if, in fact, natural forests are being replaced by planted forests.
- Forest disturbances by pests and diseases pose a significant threat, especially to new plantations. As climate variability increases, the threat to forests from fire increases as well. The moist forests of Southeast Asia, long thought to be immune to major fires, are increasingly being burned, with huge losses of timber and additional problems relating to human health and trade, for example. There is evidence that forest degradation is contributing to opening up moist tropical forests, allowing them to dry out and become more susceptible to large forest fires.
- Loss of biological diversity is a concern.
- The forest sector is witnessing a trend towards more participatory decision-making. The political commitment to sustainable forest management has never been stronger, and most countries have a relatively sound policy and legislative foundation for implementing it. There are broad trends towards more private ownership of forests, increased clarity of forest resources tenure, and decentralized management.
- One of the greatest challenges to policy-makers throughout the region is to ensure that benefits from forest products and services are shared with the poorest segments of society. Hundreds of millions of people in Asia and the Pacific continue to live below the poverty line, including in the largest countries with the fastest growing economies. A significant number of rural poor people live in forests or depend on forests in whole or in part for their livelihoods.

Problems remain to be solved, but there are increasing signs that several countries in the Asia and the Pacific region are starting to turn the corner towards sustainable forest management.

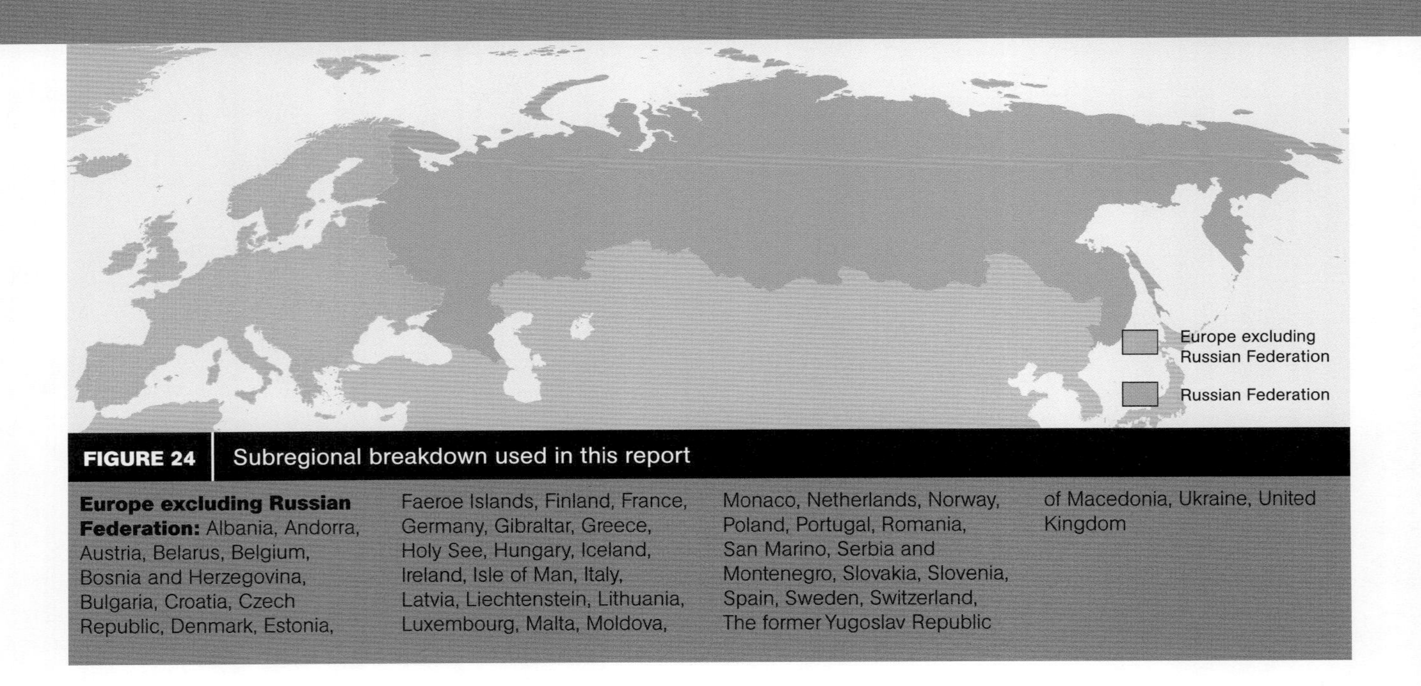


**FIGURE 24** Subregional breakdown used in this report

**Europe excluding Russian Federation:** Albania, Andorra, Austria, Belarus, Belgium, Bosnia and Herzegovina, Bulgaria, Croatia, Czech Republic, Denmark, Estonia, Faeroe Islands, Finland, France, Germany, Gibraltar, Greece, Holy See, Hungary, Iceland, Ireland, Isle of Man, Italy, Latvia, Liechtenstein, Lithuania, Luxembourg, Malta, Moldova, Monaco, Netherlands, Norway, Poland, Portugal, Romania, San Marino, Serbia and Montenegro, Slovakia, Slovenia, Spain, Sweden, Switzerland, The former Yugoslav Republic of Macedonia, Ukraine, United Kingdom

# Europe

## EXTENT OF FOREST RESOURCES

Forest statistics in Europe are dominated by the Russian Federation (including the part in Asia), which accounts for 81 percent of the total forest area. For the purposes of this study, therefore, it was decided to simply divide Europe into two categories: the Russian Federation and all other European countries.

The reported forest area for Europe in 2005 (excluding the Russian Federation) was 193 million hectares, an increase of almost 7 percent since 1990 (Figure 25 and Table 13). This compares with a net global decrease of 3 percent in forest area over the same period of time. Europe is the only major region with a net increase in forest area over the entire period of 1990–2005. (Asia has reported a net increase in the last five years, mainly a result of the massive afforestation programme in China.)

The reported net forest area in the Russian Federation is virtually stable, with a small increase in the 1990s and a small decline from 2000 to 2005.

The net increase in forest area in Europe is a result, in large part, of substantial increases in several countries over 2000–2005, led by Spain (296 000 ha/year average increase) and Italy (106 000 ha/year), followed by Bulgaria, France, Portugal and Greece. The largest percentage increases were reported by countries with low forest cover: Iceland (3.9 percent increase in forests per year) and Ireland (1.9 percent) (Figure 26).

The Russian Federation was the only European country reporting a net loss of forest area over 2000–2005, an average decrease of 96 000 ha/year; however, this amounted to only a 0.01 percent loss of total forest area.

**FIGURE 25** Extent of forest resources

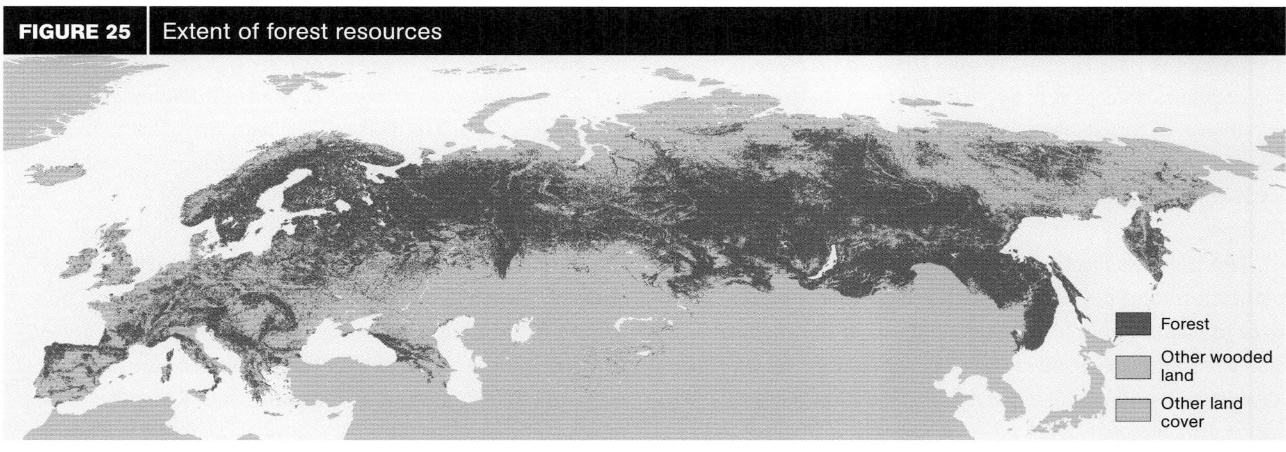


**SOURCE:** FAO, 2001a.

TABLE 13
**Extent and change of forest area**

| | Area (1 000 ha) | | | Annual change (1 000 ha) | | Annual change rate (%) | |
|---|---|---|---|---|---|---|---|
| | 1990 | 2000 | 2005 | 1990–2000 | 2000–2005 | 1990–2000 | 2000–2005 |
| Europe excluding Russian Federation | 180 370 | 188 823 | 192 604 | 845 | 756 | 0.46 | 0.40 |
| Russian Federation | 808 950 | 809 268 | 808 790 | 32 | –96 | 0 | –0.01 |
| **Total Europe** | **989 320** | **998 091** | **1 001 394** | **877** | **661** | **0.09** | **0.07** |
| **World** | **4 077 291** | **3 988 610** | **3 952 025** | **–8 868** | **–7 317** | **–0.22** | **–0.18** |

TABLE 14
**Area of forest plantations**

| | Area (1 000 ha) | | | Annual change (1 000 ha) | |
|---|---|---|---|---|---|
| | 1990 | 2000 | 2005 | 1990–2000 | 2000–2005 |
| Europe excluding Russian Federation | 8 561 | 10 032 | 10 532 | 147 | 100 |
| Russian Federation | 12 651 | 15 360 | 16 963 | 271 | 320 |
| **Total Europe** | **21 212** | **25 393** | **27 495** | **418** | **420** |
| **World** | **101 234** | **125 525** | **139 466** | **2 424** | **2 788** |

**FIGURE 26** Forest change rates by country, 2000–2005

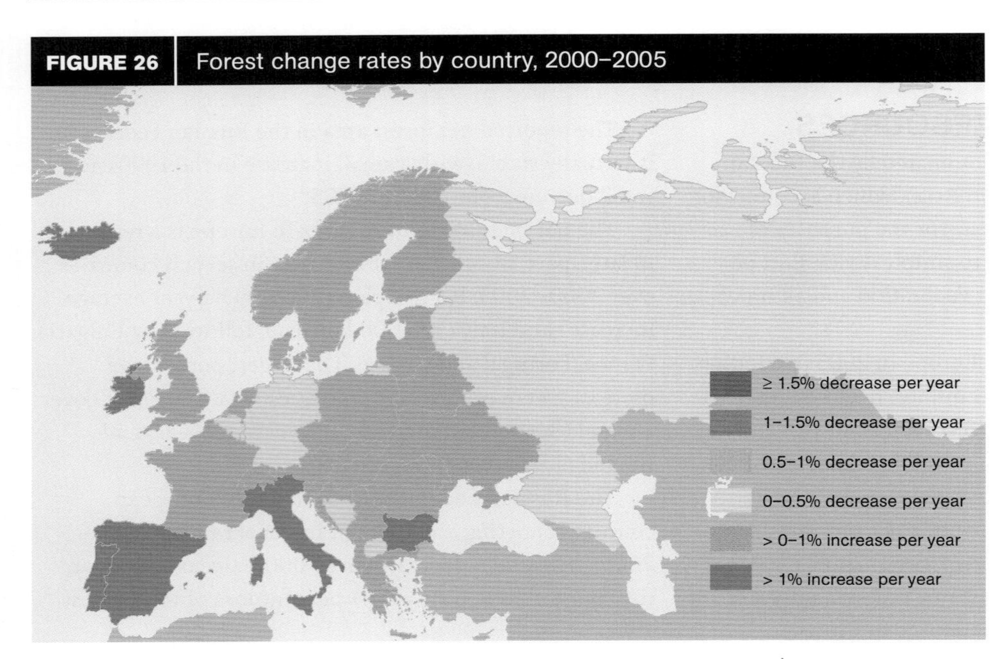


Slightly less than half of Europe's net increase in forest area over the past 15 years results from an increase in forest plantations (Table 14). The rest results from natural expansion of forests into former agricultural land and the establishment of "semi-natural" planted forests using native species, not considered to be forest plantations in Europe.

Net increases in the extent of forest, in forest plantations and in growing stock are positive trends towards sustainable forest management in the region. The Russian Federation is the only reporting country with a negative trend in this regard, but its net decrease of forest area was only 0.02 percent over the entire period of 1990–2005. All indications are that European countries have successfully stabilized or increased their forest areas, in many cases from the nineteenth or early twentieth centuries.

## BIOLOGICAL DIVERSITY

The conservation of biological diversity provides a different challenge in Europe than in other regions. While few species are currently threatened or endangered, this is mainly because much of Europe's forest has been drastically changed by human activity over several millennia. Although most of Europe has been deforested in the past under a variety of human influences such as agriculture, industrialization and war, many areas have also been reforested, naturally or intentionally, over the centuries.

Only 4 percent of Europe's forest area (excluding the Russian Federation) is classified as primary forest, compared with 27 percent of the world as a whole. The data indicate a slightly increasing trend in primary forests in Europe, other than in the Russian Federation, which accounts for 97 percent of Europe's total. Russia's primary forests increased in the 1990s, but declined by 0.2 percent per year from 2000 to 2005.

Another important proxy for conservation of biological diversity is the extent to which forest ecosystems are designated primarily for conservation. A positive global trend in the 1990s continued during 2000–2005, with the total increase over 15 years approaching 100 million hectares, an increase of 32 percent (Table 15). In Europe, the forest area designated primarily for conservation increased by 100 percent over this same period. Most of this increase occurred in the 1990s, but during 2000–2005 the increase was still significant, about 3 percent per year.

TABLE 15
**Area of forest designated primarily for conservation**

| | Area (1 000 ha) | | | Annual change (1 000 ha) | |
|---|---|---|---|---|---|
| | 1990 | 2000 | 2005 | 1990–2000 | 2000–2005 |
| Europe excluding Russian Federation | 6 588 | 17 687 | 20 272 | 1 110 | 517 |
| Russian Federation | 11 815 | 16 190 | 16 488 | 438 | 60 |
| **Total Europe** | **18 402** | **33 877** | **36 760** | **1 548** | **576** |
| **World** | **298 424** | **361 092** | **394 283** | **6 267** | **6 638** |

Some 10.5 percent of the forest area in Europe (excluding the Russian Federation) is designated for conservation, compared with a global average of 10 percent. In the Russian Federation, forest conservation area increased to 2 percent of total forest area.

The average number of threatened tree species per country in Europe is significantly less than in other regions, which would be expected, given the generally smaller number of species in these temperate and boreal ecosystems, as well as the relative stability of total forest area.

## FOREST HEALTH AND VITALITY

Fire damage to forests in the Europe region (excluding the Russian Federation) constitutes less than 10 percent of the area reported for insect pests, diseases and other disturbances. Compared with other regions of the world, non-fire disturbances are relatively well reported in Europe, with information received on over 90 percent of the forest area. However, it is difficult to compare data, since there are different interpretations of what constitutes a disturbance. Forest pests and other disturbances may have even more widespread impact than reported.

For Europe as a whole, about 2 percent of total forest area was reported affected by disturbances in a typical year (considering the annual average over 1998–2002). For Europe excluding the Russian Federation, this figure increases to about 6 percent (Table 16). Figure 27 indicates the relative disturbances caused by the four reporting categories: fire, insects, diseases and all other types (storms, drought, ice, etc.) for Europe as a whole. By far the largest disturbance in Europe was storms, which were particularly severe in 1999.

International trade has increased the risk of introduction of damaging pests and diseases. For example, *Anoplophora chinensis*, which originates in Japan and the Korean peninsula, where it is a serious pest of *Citrus* spp. and many other deciduous trees, was discovered in Europe in 2000 in Lombardy, Italy. The potential impact on the region has not yet been determined.

Within Europe, the Ministerial Conference on the Protection of Forests in Europe (MCPFE) chose defoliation as a key indicator of forest health. The International Cooperative Programme on Forests (under the United Nations Economic Commission for Europe [UNECE] Convention on Long Range Transboundary Air Pollution) has systematically monitored the crown condition of forests since the mid-1980s, when the health of Europe's forests became a matter of particular concern.

**FIGURE 27** Forest disturbances, 1998–2002

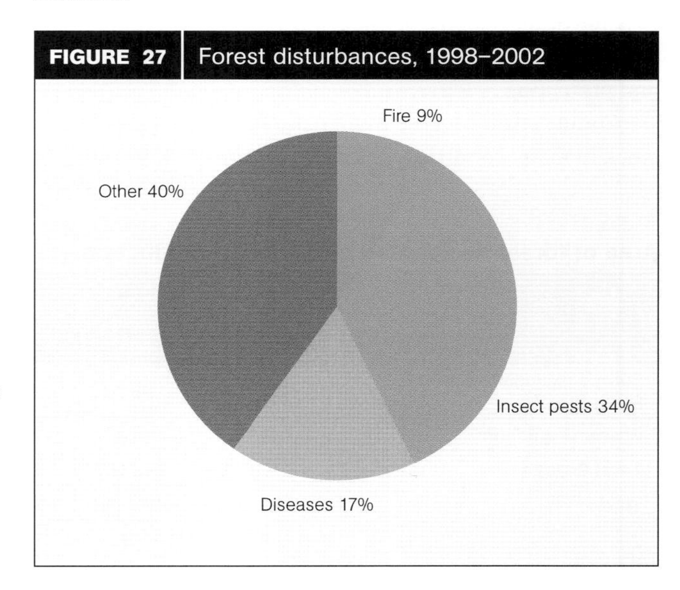


TABLE 16
**Forest disturbances**

| | Disturbances affecting forests, average 1998–2002 (1 000 ha) | | | | |
|---|---|---|---|---|---|
| | Fire | Insects | Diseases | Other | Total |
| Europe excluding Russian Federation | 326 | 1 400 | 2 178 | 7 038 | 10 942 |
| Russian Federation | 1 268 | 4 953 | 957 | 508 | 7 686 |
| **Total Europe** | **1 594** | **6 353** | **3 135** | **7 546** | **18 628** |

The thirty-third session of the European Forestry Commission (FAO, 2006f), in discussing the vulnerability of the region's forests, considered how sector policy-makers could reduce forest vulnerability to extreme climatic events, insect pests, fire, climate change and other threats. Several countries have compiled or are compiling information on their experiences in responding to disasters as a basis for future emergency action.

The absence of baseline data for earlier reporting periods makes it difficult to determine whether forest health is improving or declining. However, if from 2 to 6 percent of forest area is affected in an average year, clearly the cumulative effects and the long-term consequences, including economic impacts, can be significant.

## PRODUCTIVE FUNCTIONS OF FOREST RESOURCES

In Europe, 73 percent of total forest area is designated primarily for production (52 percent, excluding the Russian Federation), compared with a global average of 31 percent (Table 17).

The area of Europe's forests designated primarily for production declined significantly in the 1990s, but remained relatively stable during 2000–2005. The concept of forests for production is less applicable in Europe than in some other regions, because most forests in Europe are designated for multiple use, which includes production and protection.

Country data suggest an increase in the total growing stock in many countries, especially in areas of Central Europe where conservative silviculture and weak markets have brought growing stock per hectare to record high levels. The net result at the regional level is an increase both in total growing stock in cubic metres and in cubic metres per hectare (Table 18).

With the Russian Federation excluded, growing stock in Europe increased at a rate of 1.3 percent per year over 2000–2005, slightly lower than the rate of 1.4 percent in the 1990s. Growing stock also continues to increase slightly in the Russian Federation, but Russia has lower growing stock per hectare than the rest of Europe. This is to be expected, considering its vast forest areas in colder regions. The Russian Federation accounts for almost 19 percent of the world's total forest growing stock, about the same as Brazil, the other leading country in this regard.

Another indicator of the productive functions of forests is the level of wood removals. During 2000–2005, wood removals increased at about 2 percent per year for Europe as a whole. This was led by a strong rebound in the Russian Federation, where wood removals had declined sharply in the 1990s (Figure 28).

Regarding NWFPs, European countries reported removals of about 272 000 tonnes of food products from forests in 2005 (about 6 percent of the world total); 6 500 tonnes of raw material for medicine and aromatic products (5 percent); and 232 000 tonnes of other plant products (18 percent) (UNECE/FAO, 2005).

Europe's forests are among the primary producers of wood in the world. Excluding the Russian Federation, Europe accounts for 23 percent of the world's industrial roundwood removals, but only 5 percent of the world's forest area. When the Russian Federation is included,

TABLE 17
**Area of forest designated primarily for production**

| | Area (1 000 ha) | | | Annual change (1 000 ha) | |
|---|---|---|---|---|---|
| | 1990 | 2000 | 2005 | 1990–2000 | 2000–2005 |
| Europe excluding Russian Federation | 105 754 | 98 931 | 99 007 | −682 | 15 |
| Russian Federation | 664 754 | 623 120 | 622 349 | −4 163 | −154 |
| **Total Europe** | **770 508** | **722 051** | **721 355** | **−4 846** | **−139** |
| **World** | **1 324 549** | **1 281 612** | **1 256 266** | **−4 294** | **−5 069** |

TABLE 18
**Growing stock**

| | Growing stock | | | | | |
|---|---|---|---|---|---|---|
| | (million $m^3$) | | | ($m^3$/ha) | | |
| | 1990 | 2000 | 2005 | 1990 | 2000 | 2005 |
| Europe excluding Russian Federation | 22 024 | 25 103 | 26 785 | 124 | 135 | 141 |
| Russian Federation | 80 040 | 80 270 | 80 479 | 99 | 99 | 100 |
| **Total Europe** | **102 063** | **105 374** | **107 264** | **103** | **106** | **107** |
| **World** | **445 252** | **439 000** | **434 219** | **109** | **110** | **110** |

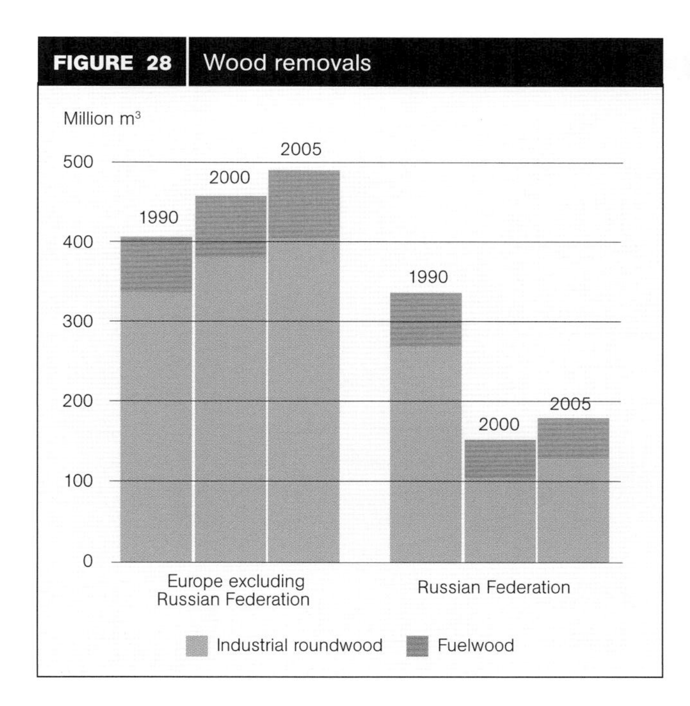


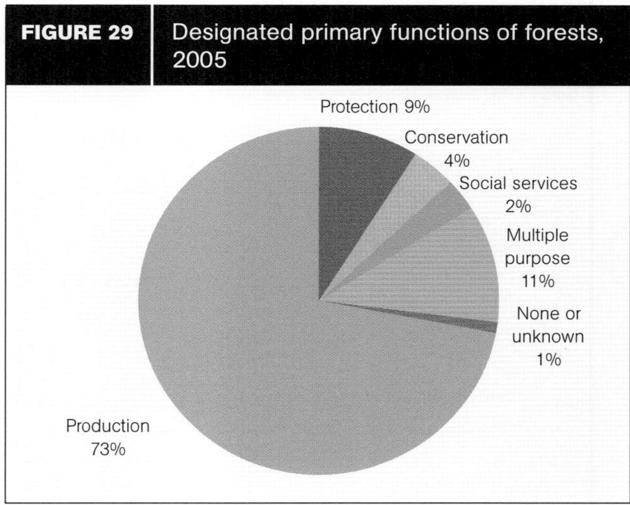


Europe accounts for 30 percent of industrial roundwood and 25 percent of forest area. Over half of Europe's forests are designated for production, much higher than the global average of 32 percent. However, as mentioned, many of the forest areas in Europe that are designated for production are also designated for other uses.

When this information is combined with the fact that Europe's forest area and growing stock continue to increase, an obvious conclusion is that a high level of wood production is not incompatible with sustainable forest management – at least not in a region of the world characterized by relatively developed economies and institutions and by relatively homogeneous (and species-poor) forests with a high proportion of commercial species. Furthermore, removals are still well below the annual increment (UNECE/FAO, 2005).

## PROTECTIVE FUNCTIONS OF FOREST RESOURCES

Forest area designated primarily for protection in 2005 accounted for 9 percent – the same as the global average (Table 19). However, not all countries use this designation, and some protective functions may be included under "multiple purpose" (Figure 29).

In Europe, protective forest plantations are increasing mainly in the Russian Federation, where they account for 30 percent of total forest plantations, compared with 9 percent in the rest of Europe. In many parts of Europe, notably in mountainous regions, protective functions are performed by existing natural or semi-natural forests.

The increasing trends in forest area designated primarily for protection and in protective forest plantations are indications that countries in Europe have recognized the importance of the protective functions of forests (in many cases for centuries). Concern about maintaining protective functions is behind the forest laws in many countries, notably in mountainous regions. Even though considerable research has been carried out on the benefits of forest protection, the fact that these benefits are not valued in the marketplace and are highly site specific continues to make it difficult to quantify them. The two parameters reported here are not sufficient to draw conclusions about the protection of air, water or soil quality in the region.

## SOCIO-ECONOMIC FUNCTIONS

Europe accounts for about 22 percent of the value of industrial roundwood removals. Its share of the global

TABLE 19
**Area of forest designated primarily for protection**

| | Area (1 000 ha) | | | Annual change (1 000 ha) | |
|---|---|---|---|---|---|
| | 1990 | 2000 | 2005 | 1990–2000 | 2000–2005 |
| Europe excluding Russian Federation | 19 010 | 19 214 | 19 543 | 20 | 66 |
| Russian Federation | 58 695 | 70 386 | 70 556 | 1 169 | 34 |
| **Total Europe** | **77 705** | **89 599** | **90 098** | **1 189** | **100** |
| **World** | **296 598** | **335 541** | **347 217** | **3 894** | **2 335** |

**FIGURE 30** Trends in value added in the forest sector, 1990–2000

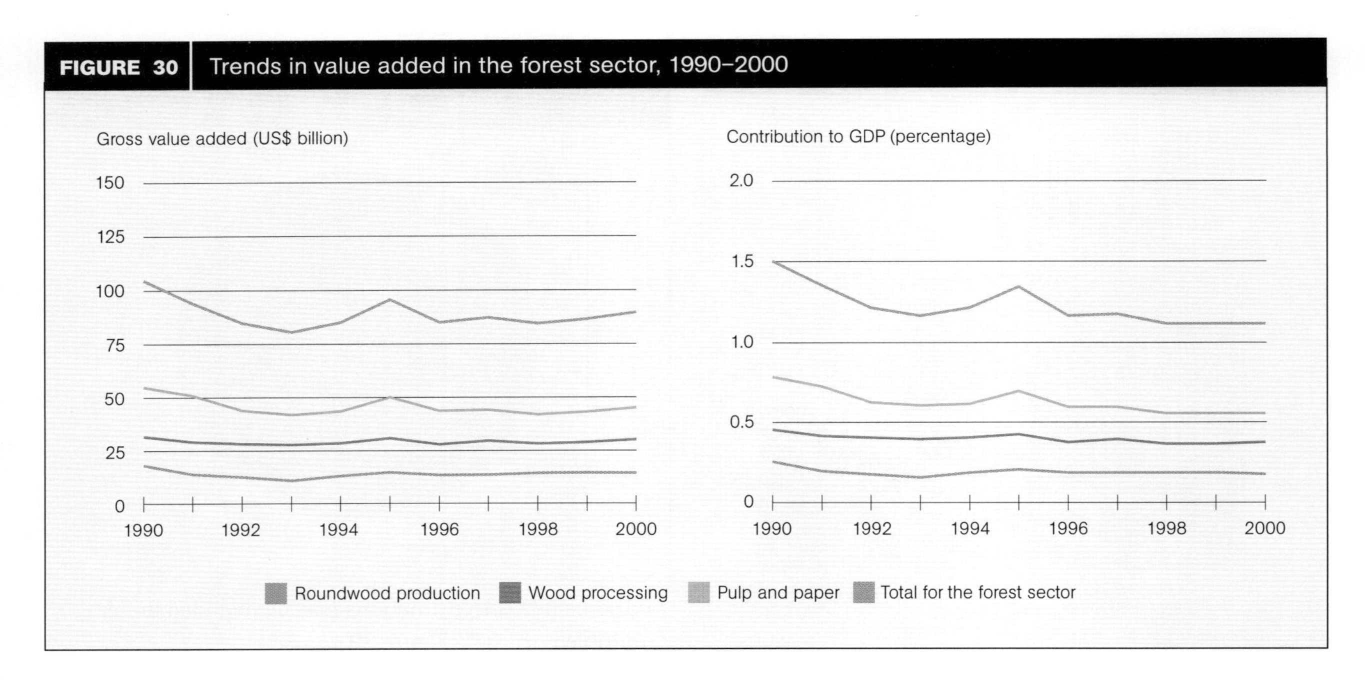


value of total wood removals has increased from 20 percent in 1990 to 22 percent in 2005. The increase has been mainly at the expense of Asia, whose share of the value of wood removals declined throughout the 1990s and continued to decline over 2000–2005.

When the net trade of forest products is considered (of both primary and secondary products), Europe leads the world as a net exporter. The sharp increase in the dollar value of European exports tends to coincide with the strengthening of the euro vis-à-vis the United States dollar.

In Europe, roundwood production accounts for only 16 percent of the total value added, compared with 34 percent for wood-processing industries and 50 percent for pulp and paper.

The data indicate a decline in value added in the forest sector in the early 1990s because of the collapse of the Russian forest sector, followed by recovery in 1995 and a levelling off in the late 1990s (Figure 30). The forest sector's contribution to GDP in Europe fell from 1.5 percent in 1990 to about 1.2 percent in 1992 and remained relatively stable thereafter.

The value of forest products trade is increasing in all parts of Europe, but the percentage increase is especially significant in Central and Eastern Europe (including European Union accession countries and other countries with economies in transition) (FAO, 2006b). The value of both exports and imports of forest products is steadily increasing.

Europe has been a net exporter of forest products since 1993 (Figure 31). Of special note are the upward trends in primary paper and wood products, and the strong surplus position in markets for secondary products. The value of exports exceeded the value of imports by US$25 billion in 2004, more than double the amount of only three years earlier.

While the increasing value of trade in forest products is impressive, it is nonetheless dwarfed by the increasing value of trade in other products and services. The value of forest products exports has declined as a share of the total value of all exports, both in Europe and in the

**FIGURE 31** Trends in net trade of forest products by subsector

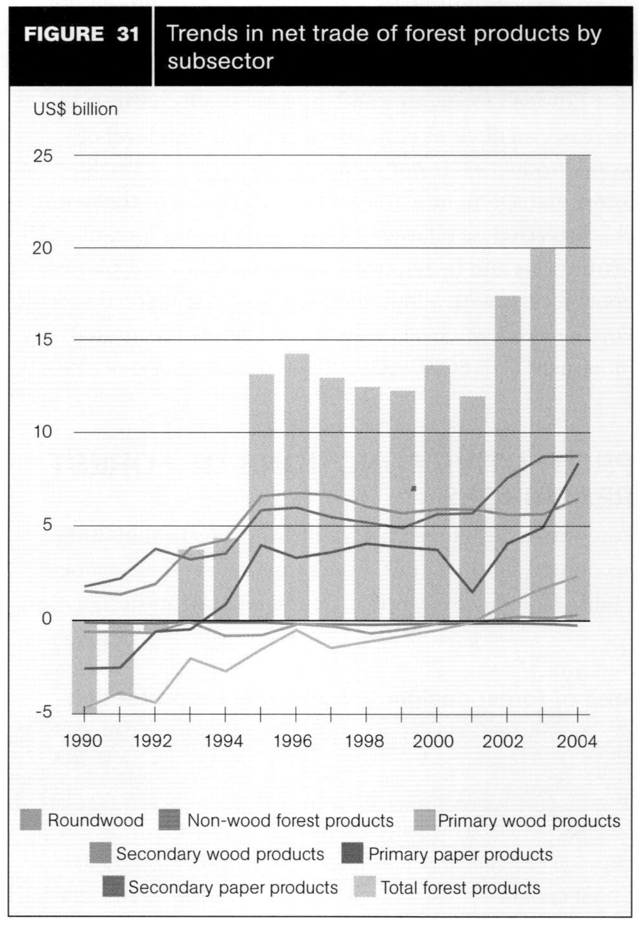


**NOTE:** A positive value indicates net export. A negative value indicates net import.

**FIGURE 32** Employment in the formal forest sector

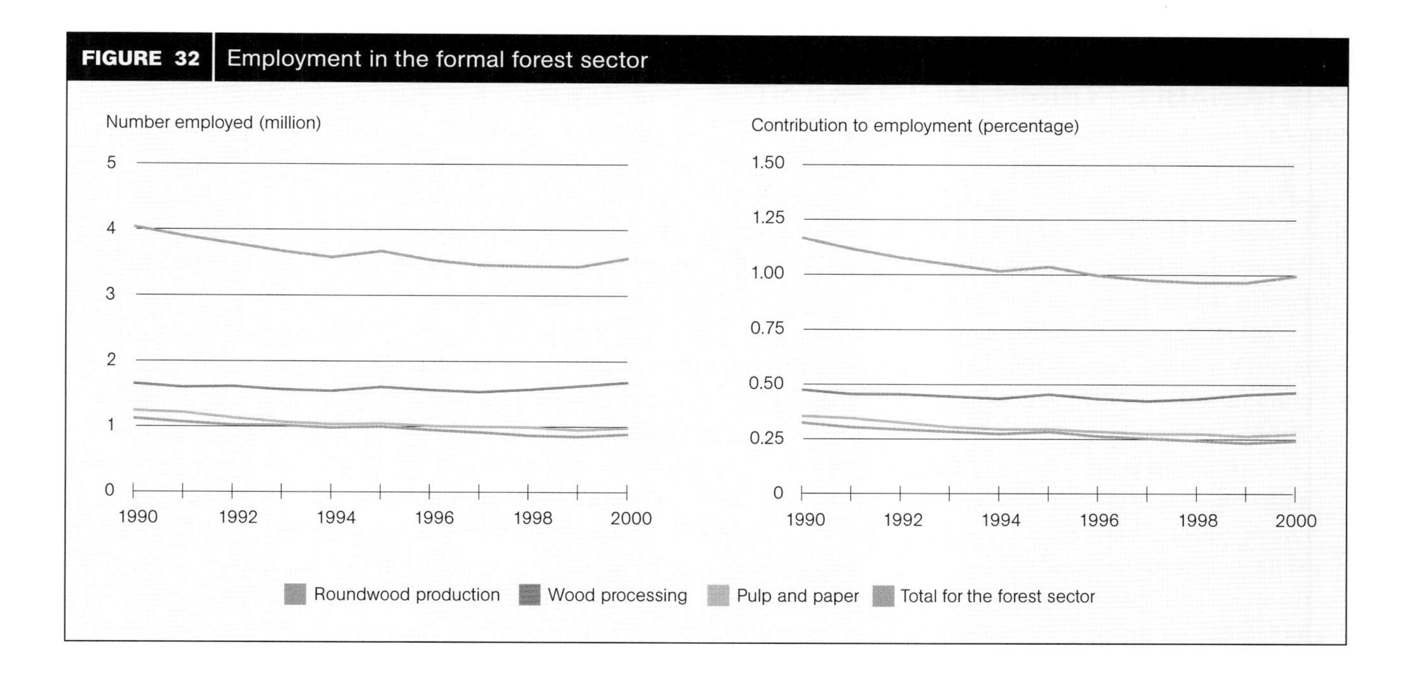


world as a whole. This decline is the most dramatic in Nordic countries, where the value of these exports increased by US$10 billion per year from 1990 to 2004. However, in percentages, this represents a decline from 21 to 13 percent of the total exports of the three Nordic countries included in this analysis (mainly resulting from the rapid rise of the telecommunications industry and other economic sectors).

Employment levels in the forest sector are declining (Figure 32), as labour productivity has been rising faster than production (UNECE/FAO, 2005).

The contribution of the forest sector to GDP has declined over the long term as other sectors, including services, have increased. As mentioned in the previous section, the marketplace tends to undervalue the protective functions of forests. However, the forest sector remains economically significant in the Baltic and Nordic countries.

## LEGAL, POLICY AND INSTITUTIONAL FRAMEWORK

"Some aspects of European forestry policy have remained remarkably stable" in the recent past (UNECE/FAO, 2005): commitment to ensuring that forest area should not decline; highly regulated forest harvesting; the requirement to replant forests after harvesting; widespread acceptance of multiple-purpose forestry practices; and tax and payment incentives that favour retention of forests and conversion of agricultural land to forest.

Forest policies are also changing in some respects, including a strong trend towards public involvement in policy-making.

In 2000, about 90 percent of Europe's forests were publicly owned and 10 percent privately owned. This statistic is heavily skewed by the Russian Federation. Excluding the Russian Federation, well over half of Europe's forests are privately owned (62 percent in the European Union).

The most important recent changes in the legal framework for forestry in Europe have taken place in Eastern Europe, where a majority of countries have reported an increase in private ownership of forests (FAO, 2006e). In several countries, the area of privately owned forests has increased by a factor of three or four in the years since the fall of the Union of Soviet Socialist Republics. However, forest ownership in the Russian Federation and the Commonwealth of Independent States remains almost 100 percent public.

An interesting trend has been the reorganizing of state forest management organizations to function as quasi-private companies, with commercial objectives and more flexibility to manage forests without following strict bureaucratic rules. Austria, Finland, Ireland, Latvia, Poland and Sweden have all made changes along these lines.

The goals of European policies, laws and institutions are remarkably similar: to promote sustainable forest management and conservation (Bauer, Kniivilä and Schmithüsen, 2004). Every country in Europe has laws and policies in place that make it very difficult to convert forests to other uses. This is true in countries where virtually all forests are owned by the state, and it is equally true in countries (mainly in Western Europe) with a large number of private forest owners.

## SUMMARY OF PROGRESS TOWARDS SUSTAINABLE FOREST MANAGEMENT

It is tempting to conclude that Europe has achieved sustainable forest management. The negative trends are largely offset by positive ones. Key indicators, including forest area, are stable or increasing, and most countries have enacted, and are capable of enforcing, laws that result in the effective protection of forests.

However, a number of disturbing trends remain:

- Forest health is adversely affected by fire, storms, insect pests and disease, all of which may increase if global warming continues.
- Climate change poses a threat to Europe's forests, although some areas may well benefit, for example from longer growing seasons.
- Employment in the forest sector continues to decline as the workforce continues to age and labour productivity increases because capital is replacing labour as the most important production factor.
- The contribution of forests to Europe's economy will most likely continue to fall if prices for forest products remain stagnant. Globalization is changing the forest sector along with the rest of the world economy.

The *European Forestry Sector Outlook Study* (UNECE/FAO, 2005) concluded that European forests are sustainable in the long term, but with caveats in all areas of sustainable development – economic, social and environmental.

Europe's leaders face many challenges, including constraints on public finances, an ageing workforce and unanswered questions about long-term economic viability – such as the impact of stagnant prices for forest products. The uncertain impact of climate change on forest ecosystems looms over Europe and the rest of the world.

However, there are also many positive trends on which to build, starting with the fact that Europe has successfully halted and reversed the historical loss of forest area. With MCPFE, Europe has in place a strong political process to support the forest sector.

**FIGURE 33** Subregional breakdown used in this report

**Caribbean:** Antigua and Barbuda, Bahamas, Barbados, Bermuda, British Virgin Islands, Cayman Islands, Cuba, Dominica, Dominican Republic, Grenada, Guadeloupe, Haiti, Jamaica, Martinique, Montserrat, Netherlands Antilles, Puerto Rico, Saint Kitts and Nevis, Saint Lucia, Saint Vincent and the Grenadines, Trinidad and Tobago, United States Virgin Islands

**Central America:** Belize, Costa Rica, El Salvador, Guatemala, Honduras, Nicaragua, Panama

**South America:** Argentina, Bolivia, Bolivarian Republic of Venezuela, Brazil, Chile, Colombia, Ecuador, Falkland Islands, French Guiana, Guyana, Paraguay, Peru, Suriname, Uruguay

**NOTE:** Mexico is included in the chapter on North America. For this reason, the regional totals for some of the data in the present chapter do not correspond to the regional totals in the *Forestry Sector Outlook Study for Latin America and the Caribbean* (FAO, 2006g).

# Latin America and the Caribbean

## EXTENT OF FOREST RESOURCES

The Latin America and the Caribbean region boasts abundant forest resources – about 47 percent of the land – and accounts for 22 percent of the world's forest area (Figure 34). The annual rate of change of forest area from 2000 to 2005 was -0.51 percent, compared with -0.46 percent during the 1990s (Table 20, Figure 35).

From 1990 to 2005, Latin America and the Caribbean lost about 64 million hectares of forest. During this period, forest area increased by 11 percent in the Caribbean and declined by 19 percent in Central America and 7 percent in South America. Forest area declined from 51 to 47 percent of the total land area in Latin America and the Caribbean during 1990–2005. The total area of other wooded land was stable, accounting for 6 percent of total land area.

Globally, forest plantations represent about 4 percent of total forest area. In the Latin America and the Caribbean region, they account for 1.4 percent of total forest area. Although this is a relatively small figure, plantations are increasing at the rate of about 1.6 percent per year (Table 21).

From 2000 to 2005, net forest area continued to decline in Central and South America. The leading cause of deforestation was the conversion of forest land to agriculture. Within the region, the largest area loss was in South America, while the largest percentage loss of forest area was in Central America. Forest area increased in Chile, Costa Rica, Cuba and Uruguay, and forest plantations increased throughout the region.

**FIGURE 34** Extent of forest resources

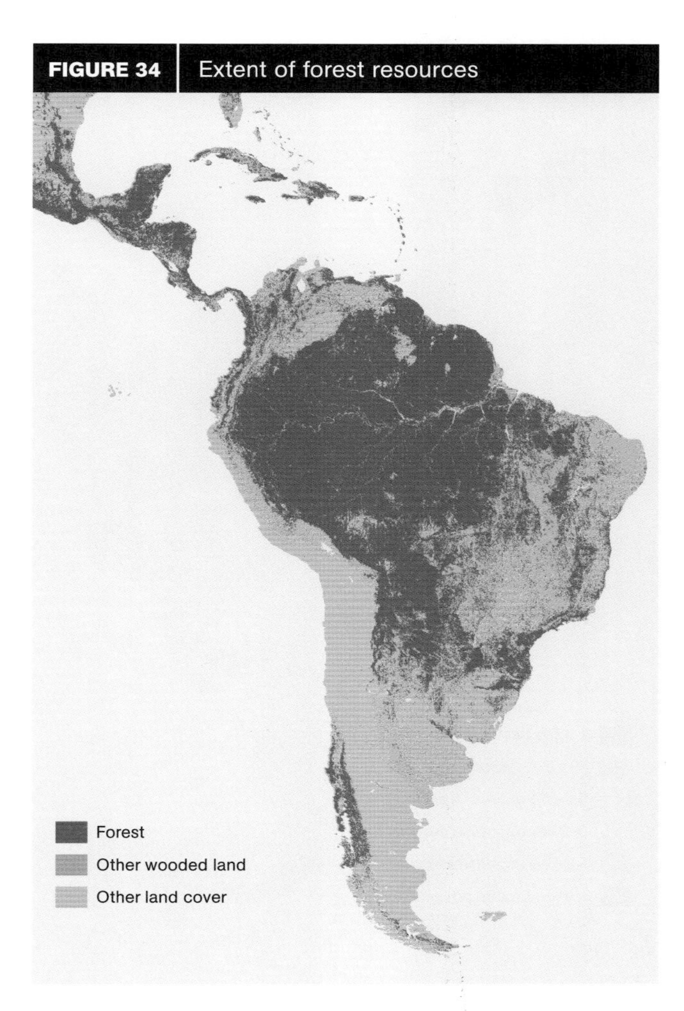


**SOURCE:** FAO, 2001a.

TABLE 20

**Extent and change of forest area**

| Subregion | Area (1 000 ha) | | | Annual change (1 000 ha) | | Annual change rate (%) | |
|---|---|---|---|---|---|---|---|
| | 1990 | 2000 | 2005 | 1990–2000 | 2000–2005 | 1990–2000 | 2000–2005 |
| Caribbean | 5 350 | 5 706 | 5974 | 36 | 54 | 0.65 | 0.92 |
| Central America | 27 639 | 23 837 | 22 411 | –380 | –285 | –1.47 | –1.23 |
| South America | 890 818 | 852 796 | 831 540 | –3 802 | –4 251 | –0.44 | –0.50 |
| **Total Latin America and the Caribbean** | **923 807** | **882 339** | **859 925** | **–4 147** | **–4 483** | **–0.46** | **–0.51** |
| **World** | **4 077 291** | **3 988 610** | **3 952 025** | **–8 868** | **–7 317** | **–0.22** | **–0.18** |

TABLE 21
**Area of forest plantations**

| Subregion | Area (1 000 ha) | | | Annual change (1 000 ha) | |
|---|---|---|---|---|---|
| | 1990 | 2000 | 2005 | 1990–2000 | 2000–2005 |
| Caribbean | 394 | 394 | 451 | 0 | 11 |
| Central America | 83 | 211 | 274 | 13 | 13 |
| South America | 8 231 | 10 574 | 11 357 | 234 | 157 |
| **Total Latin America and the Caribbean** | **8 708** | **11 180** | **12 082** | **247** | **180** |
| **World** | **101 234** | **125 525** | **139 466** | **2 424** | **2 788** |

**FIGURE 35** Forest change rates by country, 2000–2005

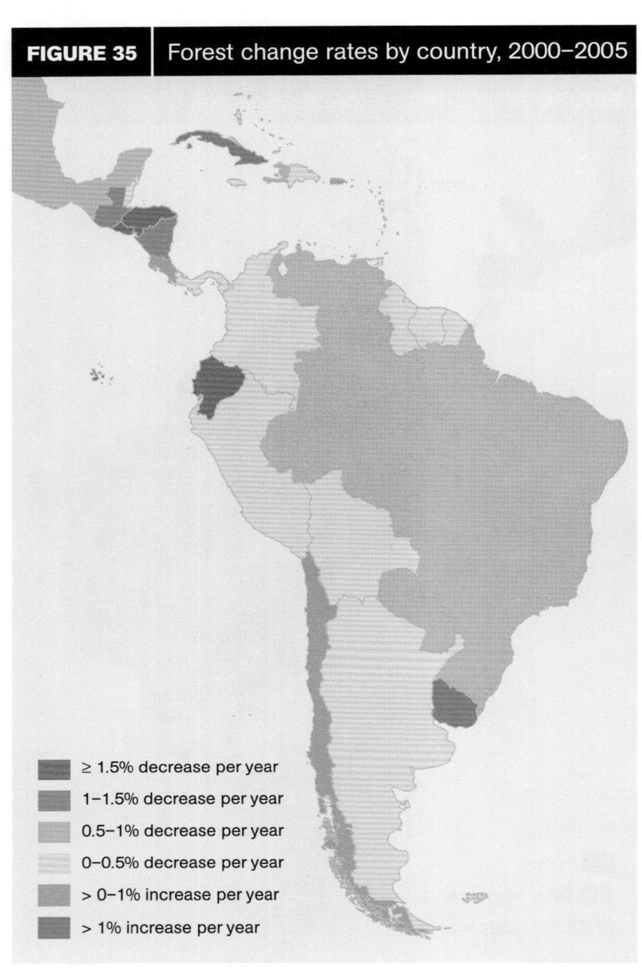


Costa Rica is an interesting, apparent success story. It is the only country in Central America that had a negative forest area change rate in the 1990s, but reported an increase in forest area from 2000 to 2005. This turnaround may be related to innovative policies for financing forest management and paying for environmental services, although macroeconomic forces causing a reduction in agricultural land may also play a part.

## BIOLOGICAL DIVERSITY

Primary forests account for 70 percent of the region's forest area and 56 percent of the world's primary forests.

Forest area designated for the conservation of biological diversity has increased dramatically over the past 15 years, including an increase of 2 percent per year from 2000 to 2005 (Table 22). This parameter has been increasing in most other regions of the world as well.

The region enjoys extremely rich forest biodiversity: no fewer than ten countries have at least 1 000 tree species (Figure 36). However, Latin America and the Caribbean also leads the world in the number of tree species considered endangered or vulnerable to extinction. For example, the region is home to *Swietenia macrophylla*, commonly known as big leaf mahogany, the tree species listed first in Appendix II of the Convention

**FIGURE 36** Number of native tree species

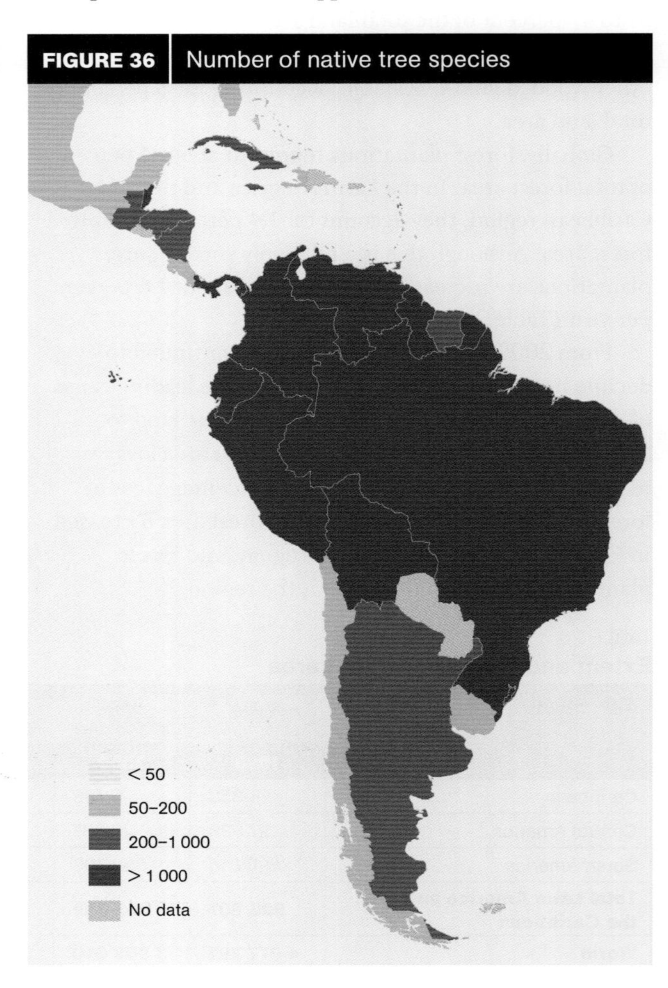

TABLE 22
**Area of forest designated primarily for conservation**

| Subregion | Area (1 000 ha) | | | Annual change (1 000 ha) | |
|---|---|---|---|---|---|
| | 1990 | 2000 | 2005 | 1990–2000 | 2000–2005 |
| Caribbean | 622 | 675 | 704 | 5 | 6 |
| Central America | 7 873 | 8 660 | 8 482 | 79 | -36 |
| South America | 69 463 | 108 103 | 119 591 | 3 864 | 2 297 |
| **Total Latin America and the Caribbean** | **77 958** | **117 439** | **128 777** | **3 948** | **2 268** |
| **World** | **298 424** | **361 092** | **394 283** | **6 267** | **6 638** |

on International Trade in Endangered Species of Wild Fauna and Flora (CITES), and which requires special trade documentation.

## FOREST HEALTH AND VITALITY

During 1999–2003, countries in South America reported an average of 26 000 wildland fires per year (FAO, 2006d), burning an average of 5.5 million hectares annually. There is a large, weather-related variation from year to year, with over 66 000 fires reported in 1997 alone, and 13.6 million hectares burned in 1999 (Figure 37). There has been an apparent long-term increase in the average number of fires and area burned, but the absence of consistent data over a long period makes it difficult to conclude with certainty that fire danger is increasing.

In the Caribbean, Cuba, the Dominican Republic and Trinidad and Tobago are the only countries that monitor fires. Their average number of fires ranged from 140 to 325 per year and average area burned ranged from 4 000 to 5 000 ha per year during 2000–2003.

In Central America, data are available for all countries except Belize. Guatemala reports the most serious wildfire problem, with an average of more than 200 000 ha burned during 2000–2003. Honduras has monitored fire activity since 1980, with an average of 2 300 fires burning 70 000 ha/year. Nicaragua reports 5 800 fires per year, affecting an average of 63 000 ha of forest land and 111 000 ha of agricultural land each year. Costa Rica averages 41 000 ha burned per year, of which about 5 000 ha are forest areas. The remaining area is divided among various other categories, for example pasture land at 15 000 ha/year.

Three subregional networks have been formed to address wildfires more effectively through the sharing of resources, expertise and information. A regional strategy for forest fire management and cooperation has been developed, and Latin America and the Caribbean has become a model for other regions considering the development of regional fire strategies.

Regarding non-fire disturbances, the southern pine beetle (*Dendroctonus frontalis*) continues to cause significant problems. It has been credited with the greatest losses of pine forests in Central America in the past 40 years (Vité *et al.*, 1975; Billings and Schmidtke, 2002), and is also the most destructive insect pest of pine forests in the southern United States of America and in parts of Mexico (Payne, 1980). The species attacks both healthy and weakened trees, such as those stressed by fire or severe weather events (for example, the outbreaks that occurred after Hurricane Mitch [1998]), and dead trees may in turn become hosts to secondary infestations and may increase fire risk. A regional bark-beetle strategy has been prepared to respond to this threat.

The introduction and subsequent establishment of forest insect pests and diseases, while having a negative impact on the forest industry in South America, has led to agreements by southern cone countries to work together to combat pests that affect regional trade. The collaboration was prompted by the discovery in Uruguay in 1986 of *Sirex noctilio*, the European woodwasp, which has now achieved widespread distribution in Argentina, Brazil and Chile, infesting several species of *Pinus* grown in commercial plantations.

**FIGURE 37** Wildland fires

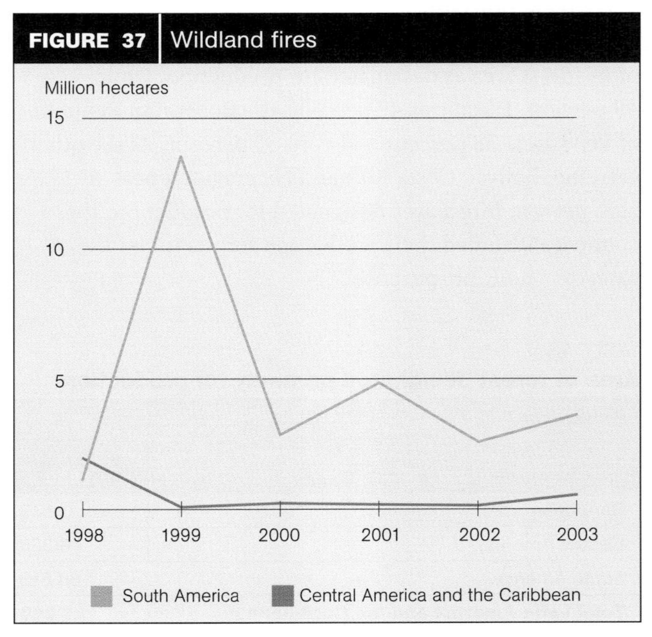


**SOURCE:** FAO, 2006d.

Examples of recent accidental introductions of forest pests into Latin America include: *Gonipterus* spp., the eucalypt snout beetle, and *Glycaspis* spp. (probably *Glycaspis brimblecombei*, the red gum psyllid), both native to Australia, which affect the growth and vigour of *Eucalyptus* spp.

Of particular significance are reports of the impact of the beaver, *Castor canadensis*, which was intentionally introduced into Argentina in 1947. The beavers are now having a significant impact on the structure of riparian forests in both Argentina and Chile. They fell many trees, and their dams cause flooding of *Nothofagus pumilio* forests, which kills the trees.

Regional plant protection organizations – including the International Plant Protection Convention (IPPC) Comité de Sanidad Vegetal del Cono Sur, Comunidad Andina, Caribbean Plant Protection Commission and Organismo Internacional Regional de Sanidad Agropecuaria – help prevent the spread and introduction of pests and promote appropriate measures for their control.

Consistent data over time are not available with enough reliability to conclude whether the long-term trend in forest health in Latin America and the Caribbean is improving or getting worse.

## PRODUCTIVE FUNCTIONS OF FOREST RESOURCES

About 12 percent of all forest area in the region is designated primarily for production, compared with a global average of 32 percent (Table 23). While the difference is significant, not all countries interpret this designation in the same way. Brazil reported only 5.5 percent of its forests in this category, bringing down the regional average. In contrast, Uruguay reports 60 percent of its forest area as designated for production, Chile 45 percent, Honduras 42 percent, the Bolivarian Republic of Venezuela 38 percent and Peru 37 percent. At the other extreme, Bolivia, Costa Rica and Nicaragua reported zero percent forest area designated for production; these countries included their production forests under the category "multiple purpose".

Growing stock as a whole is declining (Table 24), which is to be expected given declining total forest area. However, in the Caribbean it is increasing, along with forest area. Growing stock per hectare is relatively stable in Central and South America, and increasing in the Caribbean. For the region as a whole, growing stock is about 30 percent of the global total (compared with 22 percent of the global forest area), and per hectare it is 29 percent higher than the global average. According to this parameter, Latin America and the Caribbean forests are significantly more productive than those of the world at large.

In Central America and the Caribbean, by far the majority of wood removed from the forest (Figure 38) is used for fuel (90 and 82 percent, respectively).

In South America, the use of wood for fuel declined sharply in the 1990s. It continued to decline, but at a slower rate, from 2000 to 2005, while industrial roundwood continued to increase throughout the entire 15-year period. In 2005, the use of wood for industrial purposes surpassed the use of wood for fuel for the first

**FIGURE 38** Wood removals

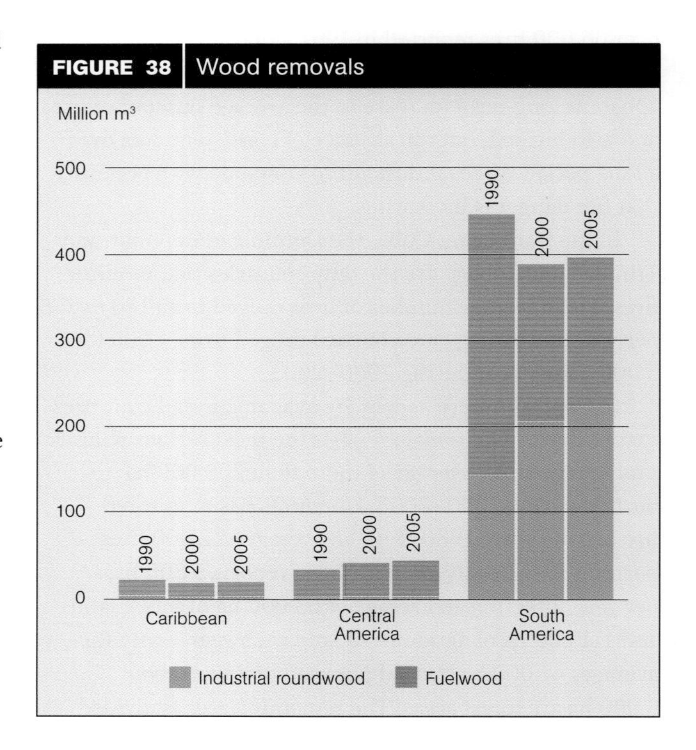


TABLE 23
**Area of forest designated primarily for production**

| Subregion | Area (1 000 ha) | | | Annual change (1 000 ha) | |
|---|---|---|---|---|---|
| | 1990 | 2000 | 2005 | 1990–2000 | 2000–2005 |
| Caribbean | 849 | 828 | 980 | −2 | 30 |
| Central America | 6 325 | 4 202 | 3 312 | −212 | −178 |
| South America | 88 216 | 103 224 | 91 073 | 1 501 | −2 430 |
| **Total Latin America and the Caribbean** | **95 390** | **108 254** | **95 364** | **1 286** | **−2 578** |
| **World** | **1 324 549** | **1 281 612** | **1 256 266** | **−4 294** | **−5 069** |

TABLE 24
**Growing stock**

| Subregion | Growing stock | | | | | |
|---|---|---|---|---|---|---|
| | *(million m³)* | | | *(m³/ha)* | | |
| | 1990 | 2000 | 2005 | 1990 | 2000 | 2005 |
| Caribbean | 328 | 403 | 441 | 61 | 71 | 74 |
| Central America | 3 585 | 3 097 | 2 906 | 130 | 130 | 130 |
| South America | 138 310 | 133 467 | 128 944 | 155 | 157 | 155 |
| **Total Latin America and the Caribbean** | **142 224** | **136 967** | **132 290** | **154** | **155** | **154** |
| **World** | **445 252** | **439 000** | **434 219** | **109** | **110** | **110** |

time. It will be interesting to see if this trend continues: there are reports that the use of wood for fuel (including biofuels for motor vehicles) is increasing in response to the rising cost of fossil fuels.

NWFPs are also significant, but the lack of available data at the regional level is such that it is not possible to draw meaningful conclusions regarding trends.

## PROTECTIVE FUNCTIONS OF FOREST RESOURCES

The regional trend in area of forest designated primarily for protective functions has been fairly stable over the past five years, after increasing in the 1990s (Table 25). Only the Caribbean showed an increase from 2000 to 2005. Forest area designated for protection represents 11 percent of the forest area in the region, compared with 9 percent globally. Several countries in Latin America and the Caribbean are among the world leaders in exploring innovative approaches to payment for environmental services such as clean water.

Protective function is another parameter that needs to be treated with caution, because many countries do not use this designation, and some protective functions may be included under "multiple purpose" (Figure 39). For example, Bolivia, the Bolivarian Republic of Venezuela, the Dominican Republic, Guatemala and Nicaragua are among the countries that did not report any forest area with this designation, and Costa Rica included only forest plantations. Brazil reported 18 percent, representing the bulk of total area designated for protection in the region.

**FIGURE 39** Designated primary functions of forests, 2005

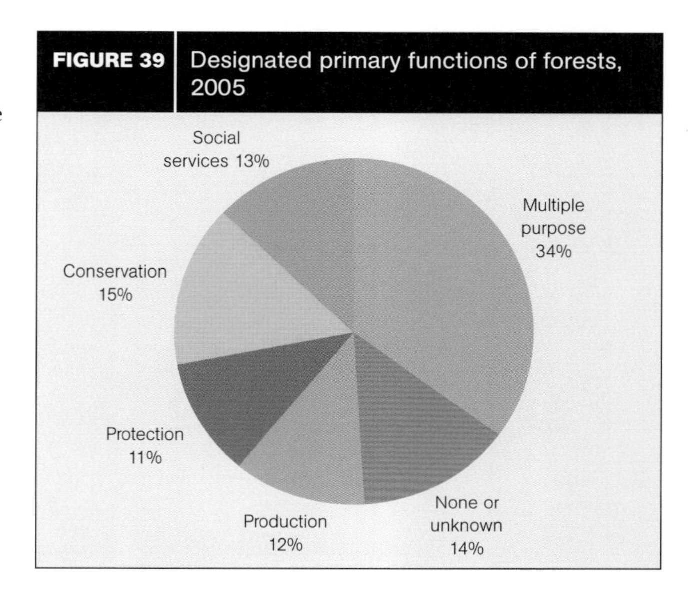


Most countries reported only a small area of forest plantations designated primarily for protection.

## SOCIO-ECONOMIC FUNCTIONS

Latin America and the Caribbean represents over 20 percent of the global forest area, but only about 7 percent of the global forest sector's value. Latin America and the Caribbean countries account for 18 percent of the value added in the primary forest sector (roundwood production), but only 3 percent of the value added in

TABLE 25
**Area of forest designated primarily for protection**

| Subregion | Area (1 000 ha) | | | Annual change (1 000 ha) | |
|---|---|---|---|---|---|
| | 1990 | 2000 | 2005 | 1990–2000 | 2000–2005 |
| Caribbean | 850 | 1 085 | 1 291 | 24 | 41 |
| Central America | 1 344 | 1 178 | 1 068 | −17 | −22 |
| South America | 90 631 | 93 632 | 93 559 | 300 | −15 |
| **Total Latin America and the Caribbean** | **92 825** | **95 895** | **95 917** | **307** | **5** |
| **World** | **296 598** | **335 541** | **347 217** | **3 894** | **2 335** |

wood-processing industries and 6 percent in the pulp-and-paper industry. This is an indication that the Latin America and the Caribbean region is a major source of raw materials, but that much of the processing of these materials into finished products is done in other regions. It is also interesting to note that the contribution of the overall forest sector to GDP is higher in Latin America and the Caribbean than in any other major region of the world.

During the 1990s, the value added by the forest sector in Latin America and the Caribbean tended to increase, but the relative contribution of the forest sector to GDP tended to decline, because other sectors were growing faster than the forest sector (Figure 40). However, the trend was reversed in 1999 and 2000, when the contribution of the forest sector to national GDP increased.

The value of forest products trade between countries has increased significantly since 1990 (Figures 41 and 42).

**FIGURE 40** Trends in value added in the forest sector, 1990–2000

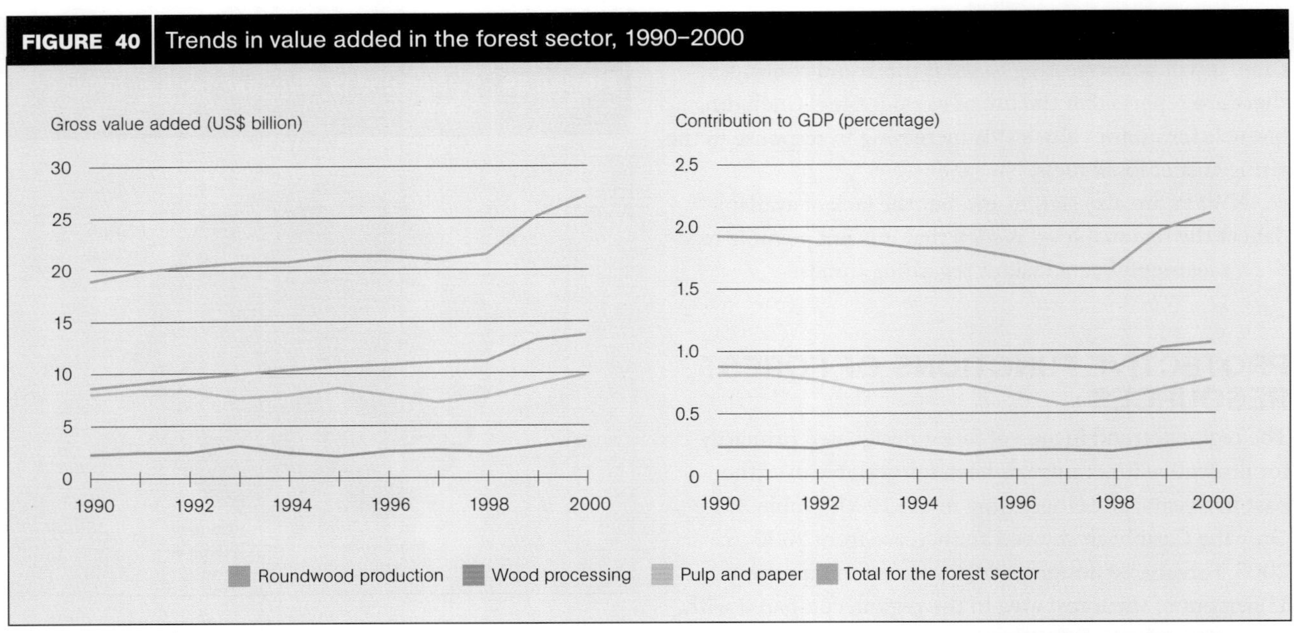


**FIGURE 41** Trends in net trade of forest products by subsector

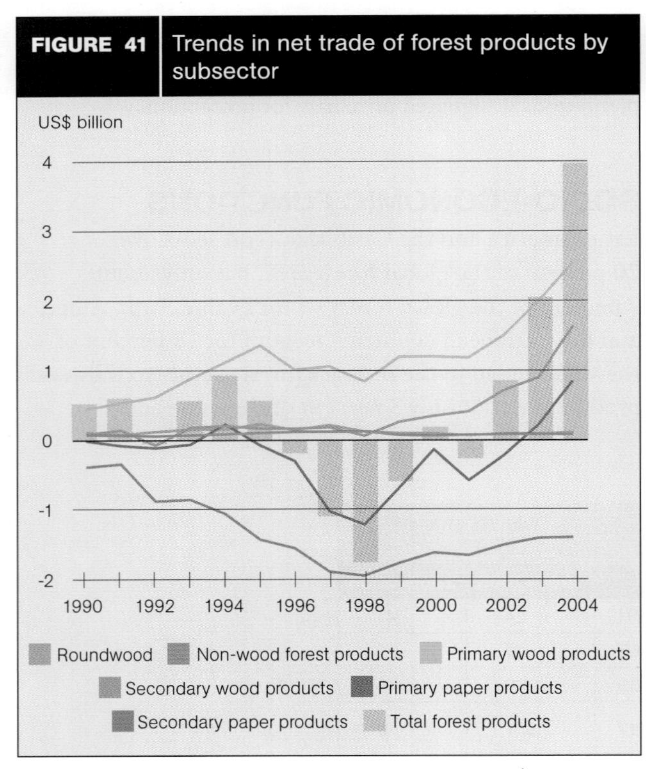

**NOTE:** A positive value indicates net export. A negative value indicates net import.

**FIGURE 42** Trends in net trade of forest products by subregion

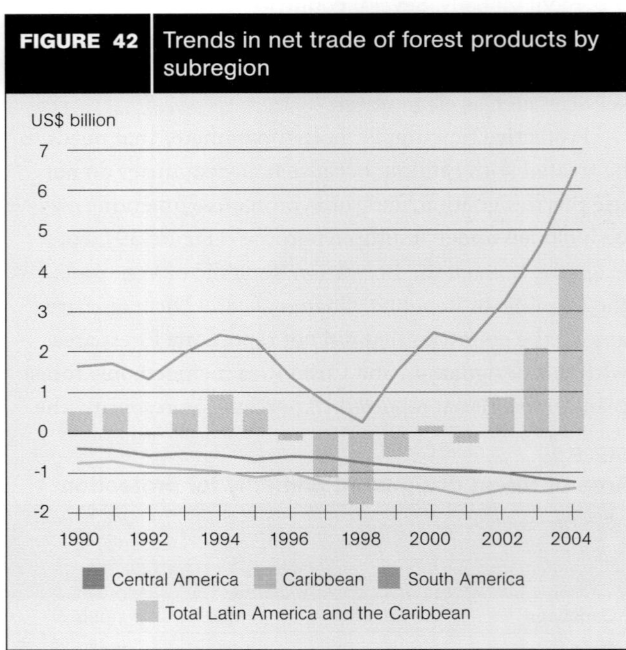


**NOTE:** A positive value indicates net export. A negative value indicates net import.

Exports have tripled in value for the region as a whole, mainly in South America. However, the import of forest products greatly exceeds exports in the Caribbean and Central America.

For Latin America and the Caribbean as a whole, the share of forest products exports as a percentage of total trade has continued to increase, from 3.7 percent in 1990 to 4.7 percent in 2004. Imports into the region accounted for 3.7 percent of total imports, equal to the global average.

Latin America and the Caribbean faced a declining trade balance in forest products from 1994 to 1998, followed by a positive trend from 1999 to 2004 led by strong increases in the export of primary and secondary wood products. The fact that the region is a net importer of secondary paper products suggests that there is potential to invest in the secondary paper industry.

Employment is another important socio-economic indicator. In much of the world, the 1990s was a period of declining forest-sector employment, but in Latin America and the Caribbean there was an upward trend from 1993 to 2000 (Figure 43). The percentage of forest-sector employment to total employment also increased from 1995 to 2000.

The Latin America and the Caribbean region's percentage of the world's forest area and growing stock are some three times higher than key economic indicators for the region, such as value of wood removals or value added. This suggests that the region has an underutilized potential for increased forest production. Some observers have suggested that a higher rate of economic development in the forest sector would lead to increased deforestation. However, a strong forest sector in terms of economic activity does not imply deforestation. On the contrary, the regions where the market value of forest products is high are the regions in which forest area is stable or increasing, i.e. Europe and North America.

In summary, the Latin America and the Caribbean region has several positive socio-economic indicators. At the end of the 1990s, the last period for which good data are available, value added and employment in the forest sector were both increasing. Subsequently, exports of forest products have continued to grow at a higher rate than imports, leading to a strong positive trade balance for the region as a whole (despite a negative trade balance in the Caribbean and Central America). In general, the economic situation in the region has more positive than negative trends.

## LEGAL, POLICY AND INSTITUTIONAL FRAMEWORK

Throughout the region, there is considerable evidence of increasing political commitment to achieve sustainable forest management. First and foremost, a majority of countries have enacted new forest laws or policies in the past 15 years, or have taken steps to strengthen existing legislation or policies. Among countries that have enacted new forest legislation (FAO, 2006e) are:

- Caribbean: Cuba, Dominican Republic, Jamaica, Saint Vincent and the Grenadines;
- Central America: Costa Rica, El Salvador, Guatemala, Honduras, Nicaragua, Panama;
- South America: Argentina, Bolivia, Brazil, Chile, Colombia, Ecuador, Paraguay, Peru, Suriname.

Latin America has a number of active regional processes that promote collaboration among members, including the Amazon Treaty Cooperation Organization,

**FIGURE 43** Employment in the formal forest sector

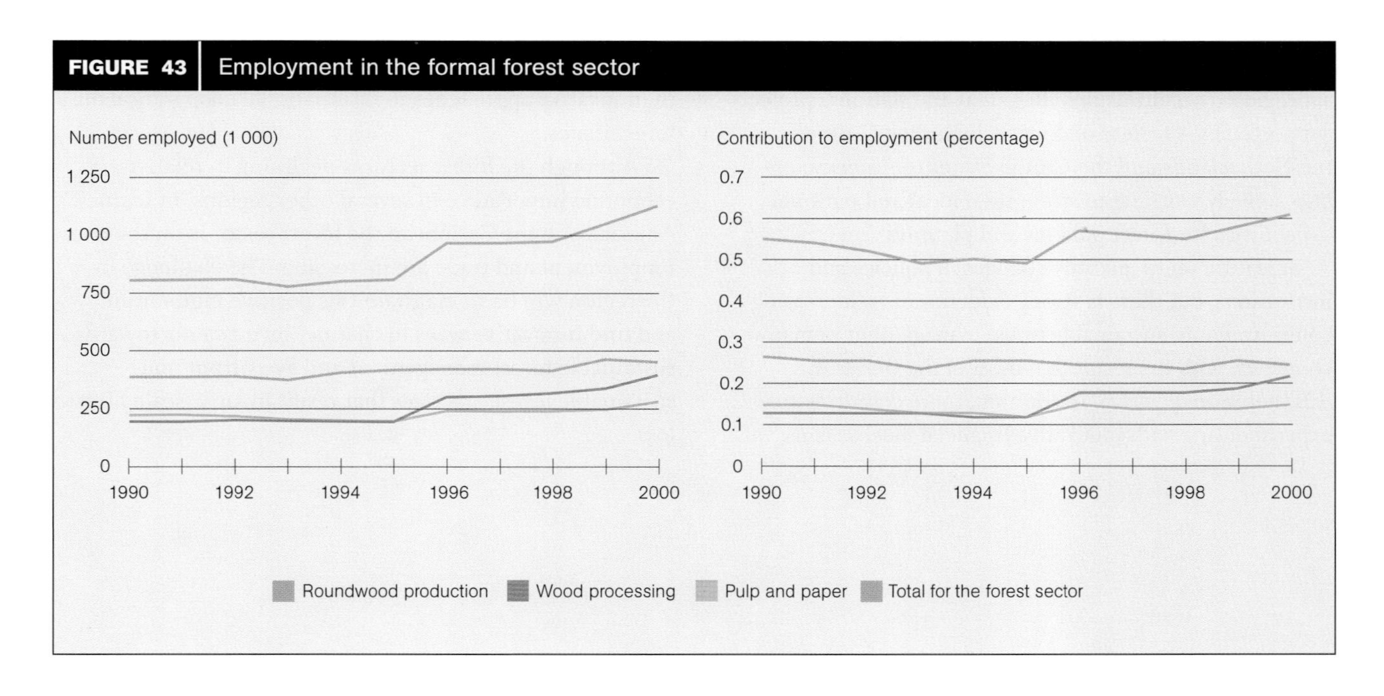

the Central American Commission on Environment and Development, the Caribbean Natural Resources Institute and the Centro Agronómico Tropical de Investigación y Enseñeanza.

Active regional networks in Latin America and the Caribbean include:

- fire management networks in South America, Central America and the Caribbean;
- a regional technical cooperation network in watershed management (Red Latinoamericana de Cooperación Técnica en Manejo de Cuencas Hidrográficas – REDLACH);
- a regional technical cooperation network in national parks and protected flora and fauna and other protected areas (Red Latinoamericana de Cooperación Técnica en Parques Nacionales, otras Áreas Protegidas, Flora y Fauna Silvestres – REDPARQUES);
- a network of national forest programme focal points.

Countries also participate in several processes that promote the use of criteria and indicators for sustainable forest management, including the Lepaterique, Montreal and Tarapoto processes and the criteria and indicators of the International Tropical Timber Organization (ITTO).

Several forestry administrations are integrating planning processes and policies so that forest-sector planning is less isolated from other planning processes, including a greater focus on forest-sector development. A number of countries are also moving towards decentralization of activities and policies affecting forests, including more integration with other sectors, development of new national forest plans with broader stakeholder participation and more emphasis on effective forest law enforcement.

The National Forest Programme Facility is supporting participatory forest processes through grants to more than 50 local and national NGOs in nine countries and three subregional organizations. Regional and national projects supported by a variety of donors, including Germany, the Netherlands and the United States of America, are also actively working to strengthen local and national capabilities for forest policies and planning.

It is difficult to quantify changes in policies and institutions, but there is strong evidence of an increasing commitment to sustainable forest management in many countries, with the result that most of the trends in this theme are positive. In addition, more countries are experimenting with innovative financial mechanisms, decentralization and participatory processes in the forest sector.

## SUMMARY OF PROGRESS TOWARDS SUSTAINABLE FOREST MANAGEMENT

The continuing high rates of conversion from forests to other land uses in many countries in Latin America and the Caribbean is a matter of great concern to decision-makers in the region as well as to external observers. Macroeconomic forces resulting in lower market prices for forest products than for products of other sectors make it difficult to manage forests with a long-term perspective.

A limiting factor for some countries that are striving to improve forest management is a shortage of financial resources. Most forests in the region are publicly owned, but public resources are increasingly scarce, or the allocated share of the public budget is inadequate. The forest sector must do a better job of making the benefits of forests known to political decision-makers, as well as promoting sustainable private-sector investment in forests.

Costa Rica is an apparent success story. It is the only country in the region that reported a negative forest area change rate in the 1990s and an increase in forest area from 2000 to 2005. It is not clear to what extent this turnaround is related to a reduction in agricultural land or to innovative policies.

The positive trend in forest area in the Caribbean is also very encouraging, although the lack of good information about forest resources is such that not too many conclusions can be drawn about trends, especially among some of the smaller island states.

The large increase in forest area designated for biodiversity conservation is a positive trend indicating that countries are taking steps to try to stop the loss of primary forests. In addition, the region is among the world leaders in innovative approaches to international cooperation on forest issues.

Although the forest sector is declining in relative economic importance in several other regions, in Latin America and the Caribbean the forest sector is on the rise. Employment and trade are increasing. The challenge in the region will be to maintain this positive momentum and find innovative ways to channel investments towards sustainable forest management and away from non-sustainable forest practices that result in large-scale forest loss.

## BOX 3 ITTO study on the state of tropical forest management

*Status of tropical forest management 2005* (ITTO, 2006) evaluates forest status in producer member countries of the International Tropical Timber Organization.[1] It complements the Global Forest Resources Assessment 2005 (FAO, 2006a) and the *Annual review and assessment of the world timber situation 2005* (ITTO, 2005). Together they provide a comprehensive picture of the changing condition of the world's forests.

A 1988 survey by ITTO found that less than 1 million hectares of tropical forest was being managed in accordance with good forestry practices. The new ITTO report considers changes in the subsequent 17 years in the 33 ITTO member countries, which are producers of tropical timber.

ITTO urges countries to undertake land-use planning, in which land is assigned as "permanent forest estate" for the sustainable production of timber and other forest goods and services. On this land, ITTO encourages countries to adopt sustainable forest management, through which the inherent values of the forest are maintained (or at least not unduly reduced), while revenues are earned, people are employed and communities are sustained by the production of timber and other forest products and services. ITTO also developed criteria and indicators for the monitoring, assessment and reporting of sustainable forest management, and these formed the basis of the assessments carried out for the report.

The forest management situation was analysed in all 33 producer member countries. The results are summarized by region (table).

**Results**

Despite difficulties and some notable deficiencies, the report found that there has been significant progress towards sustainable forest management in the tropics since 1988. Countries have established and are starting to implement new forest policies that contain the basic elements of such management. More forests have been given some security by commitment as permanent forest estate, or a similar concept, for production or protection, and more are actually being managed sustainably. Moreover, some of the permanent forest estate is certified – a new development since 1988. This is encouraging, but the proportion of natural, production forest under sustainable management is still very low and is distributed unevenly across the tropics and within countries.

At present, the natural permanent forest estate in Africa, Asia and the Pacific, and Latin America and the Caribbean is estimated to cover 110, 168 and 536 million hectares respectively, a total of 814 million hectares in the 33 ITTO member producer countries. Of the permanent forest estate in Latin America and the Caribbean, nearly half (271 million hectares) is made up of protection permanent forest estate in Brazil. Estimates of total forest area vary according to source. At the high end of the range of estimates, Africa has 274 million hectares of forest (40 percent of which is in the permanent forest estate); at the low end, 234 million hectares (47 percent of which is in it). In Asia and the Pacific, the figures are 316 million hectares (65 percent) and 283 million hectares (73 percent), respectively; in Latin America and the Caribbean, they are 931 million hectares (58 percent) and 766 million hectares (71 percent).

**Management status in the tropical permanent forest estate *(1 000 ha)***

| Region | Production | | | | | | | Protection | | | All | |
|---|---|---|---|---|---|---|---|---|---|---|---|---|
| | Natural | | | | Planted | | | | | | | |
| | Total area | With management plans | Certified | Sustainably managed | Total area | With management plans | Certified | Total area | With management plans | Sustainably managed | Total area | Sustainably managed |
| Africa | 70 461 | 10 016 | 1 480 | 4 303 | 825 | 488 | 0 | 39 271 | 1 216 | 1 728 | 110 557 | 6 031 |
| Asia and the Pacific | 97 377 | 55 060 | 4 914 | 14 397 | 38 349 | 11 456 | 184 | 70 979 | 8 247 | 5 147 | 206 705 | 19 544 |
| Latin America and the Caribbean | 184 727 | 31 174 | 4 150 | 6 468 | 5 604 | 2 371 | 1 589 | 351 249 | 8 374 | 4 343 | 541 580 | 10 811 |
| **Total** | **352 565** | **96 250** | **10 544** | **25 168** | **44 778** | **14 315** | **1 773** | **461 499** | **17 837** | **11 218** | **858 842** | **36 386** |

[1] Africa – Cameroon, Central African Republic, Congo, Côte d'Ivoire, Democratic Republic of the Congo, Gabon, Ghana, Liberia, Nigeria and Togo; Asia and the Pacific – Cambodia, Fiji, India, Indonesia, Malaysia, Myanmar, Philippines, Papua New Guinea, Thailand and Vanuatu; and Latin America and the Caribbean – Bolivia, Bolivarian Republic of Venezuela, Brazil, Colombia, Ecuador, Guatemala, Guyana, Honduras, Mexico, Panama, Peru, Suriname and Trinidad and Tobago.

It is always possible for a country to unprotect areas of permanent forest estate for purposes that it considers important. Some countries have still not clearly identified one (some have not even adopted the term or a concept equivalent to it), and some have undergone political changes that have acted to obfuscate forest ownership. And there are still frequent conflicts over tenure that engage governments, local communities and private owners – issues that must be resolved if the forest is to be rendered more secure. Taking the tropics as a whole, however, there has been great improvement in the legal security of both production and protection forests in the last two decades. In addition, security has now been increased in many countries through better delimitation of boundaries.

The area of natural, production permanent forest estate in ITTO producer member countries is estimated at 353 million hectares. Of this, an estimated 96.3 million hectares (27 percent of the total) are covered by management plans, 10.5 million hectares (3.0 percent) are certified by an independent certification organization, and at least 25.2 million hectares (7.1 percent) are managed sustainably. The area of protection permanent forest estate in ITTO producer member countries is estimated to be 461 million hectares, of which an estimated 17.8 million hectares (3.9 percent) are covered by management plans and at least 11.2 million hectares (2.4 percent) are being managed sustainably. A much larger but unestimated area of the forest estate is not under immediate threat from anthropogenic destructive agents, being remote from large human settlements and projected roads.

Thus the proportion of the tropical, production permanent forest estate managed sustainably has grown substantially since 1988, from less than 1 million hectares to more than 25 million – and to more than 36 million hectares if the area of protection permanent forest estate so managed is included. Despite this significant improvement, the overall proportion known to be sustainably managed remains very low, at less than 5 percent of the total.

### Constraints to sustainable forest management

The report identified several constraints on the diffusion of sustainable forest management. Probably the most important, and the most generally applicable, is that sustainable management for the production of timber is less profitable to landowners and users than many other possible ways of using the land.

Another constraint is related to land tenure. There have been advances in many countries in committing forest for either production or protection and in establishing a permanent forest estate. However, without the security provided by long-term government resolve and credible arrangements for tenure, sustainable forest management is unlikely to succeed.

Illegal logging and the illegal movement of timber have become pressing issues in many countries, exacerbated by local warfare and by drug smuggling and other criminal activities. These have not only made forest management in the field a hazardous business and prejudiced the security of many permanent forest estates, but they have also undermined legitimate markets for timber and reduced the profitability of legitimate producers.

There is an almost universal lack of the resources needed to manage tropical forests properly. There are chronic shortages of staff, equipment, vehicles and facilities for research and training. The pay and conditions of service are rarely sufficiently favourable to attract and keep skilled staff in the field.

In the preparation of the report it became clear that, in most countries, information on the extent of forests and the status of management in the permanent forest estate is still very poor. The report should encourage ITTO member countries and forest-related institutions and organizations to continue to improve their data-collection systems, as reliable information is the cornerstone of both practising and assessing sustainable forest management.

### Conclusion and recommendations

Despite the progress made since 1988, significant areas of tropical forest are still lost every year, and unsustainable (and often illegal) extraction of tropical forest resources remains widespread. However, with most countries now attempting widespread implementation of sustainable forest management, it is hoped that the pace of progress will accelerate in coming years.

The report made three recommendations to help quicken the pace: that regular reporting on the status of tropical forest management be instituted at the international level; that the international community make resources available to improve the capacity of countries to collect, analyse and make available comprehensive data on the status of tropical forest management; and that the international forest-related community set as its first priority the development of a system for ensuring that sustainable forest management is a financially remunerative land use.

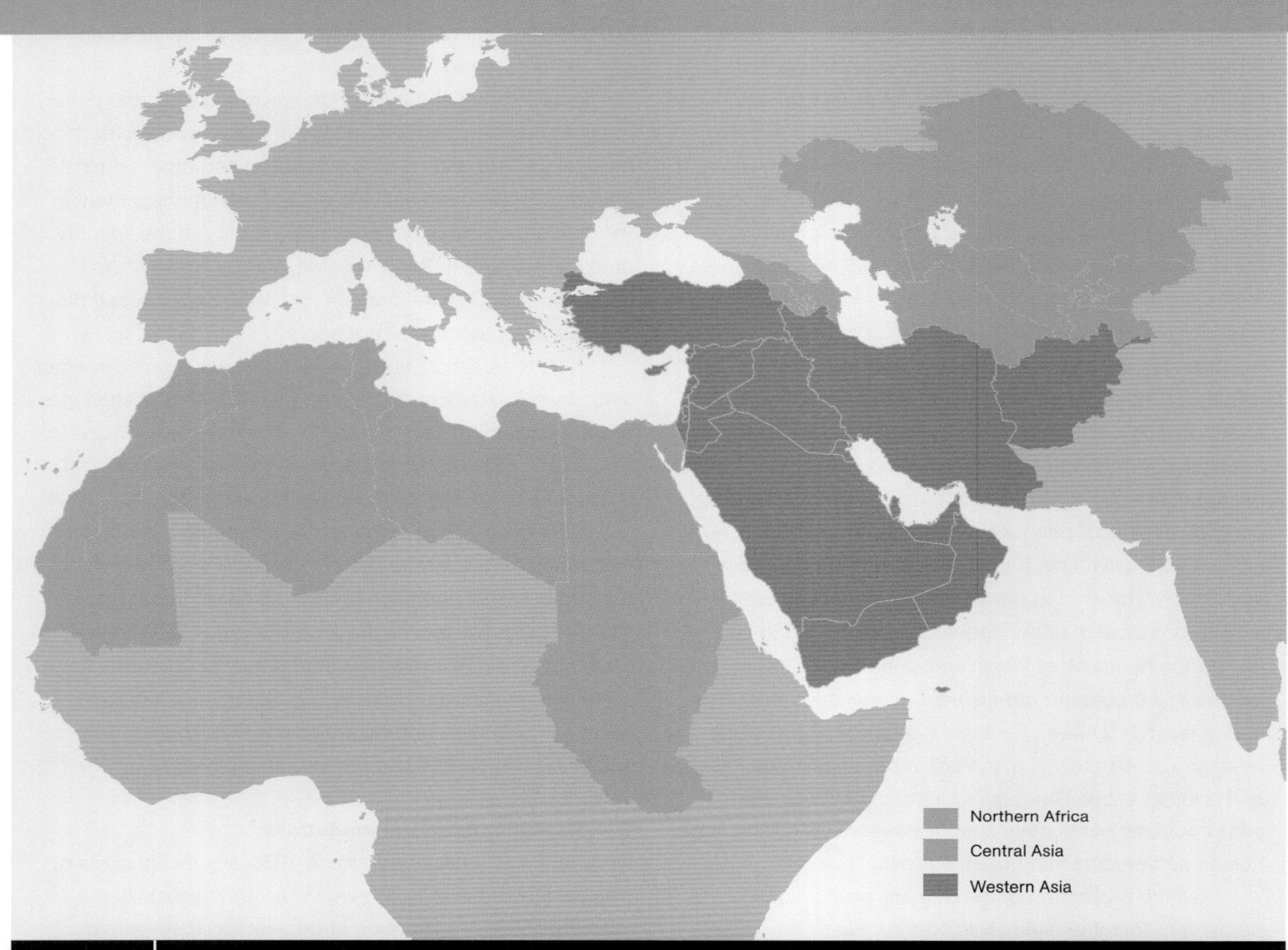


**FIGURE 44** Subregional breakdown used in this report

**Northern Africa**: Algeria, Egypt, Libyan Arab Jamahiriya, Mauritania, Morocco, Sudan, Tunisia

**Central Asia**: Armenia, Azerbaijan, Georgia, Kazakhstan, Kyrgyzstan, Tajikistan, Turkmenistan, Uzbekistan

**Western Asia**: Afghanistan, Bahrain, Cyprus, Islamic Republic of Iran, Iraq, Israel, Jordan, Kuwait, Lebanon, Oman, Qatar, Saudi Arabia, Syrian Arab Republic, Turkey, United Arab Emirates, Yemen

# Near East

The present report divides the Near East region into three areas based on geographic proximity and similar forest ecological characteristics: Northern Africa, Central Asia and Western Asia (Figure 44).

The countries in Northern Africa are also included in the chapter on Africa. Hence, the totals found in the regional tables throughout this report should not be summed, as it would result in double counting. Global statistics can be found in the tables in the annex of this report or in the main report of FRA 2005 (FAO, 2006a).

## EXTENT OF FOREST RESOURCES

The extent of forest resources in the Near East is very low. The estimated forest area for the region in FRA 2005 was 120 million hectares, about 3 percent of the world's forest area (Figure 45 and Table 26). In contrast, the Near East has 15 percent of the world's land area. Forests cover about 6 percent of the land area in the Near East, compared with 30 percent globally. Of the 31 countries included in this report, seven have forest cover exceeding 10 percent of the total land area: Armenia, Azerbaijan, Cyprus, Georgia, Lebanon, the Sudan and Turkey. The remaining 24 countries are considered to have low forest cover (less than 10 percent of the land area).

The world lost about 3 percent of its forest area from 1990 to 2005. In Central and Western Asia, forest area is essentially stable – it is declining slightly in some countries and increasing slightly in others, with the exception of Afghanistan, where it is declining rapidly (Figure 46).

**FIGURE 45** Extent of forest resources

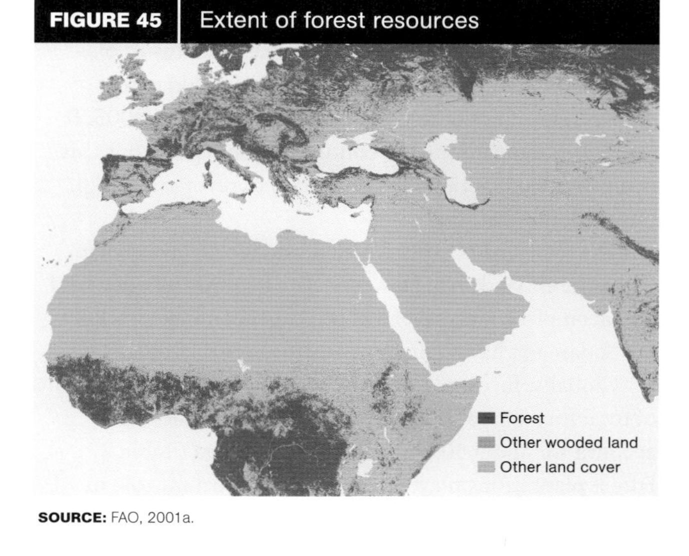


**SOURCE:** FAO, 2001a.

**FIGURE 46** Forest change rates by country or reporting area, 2000–2005

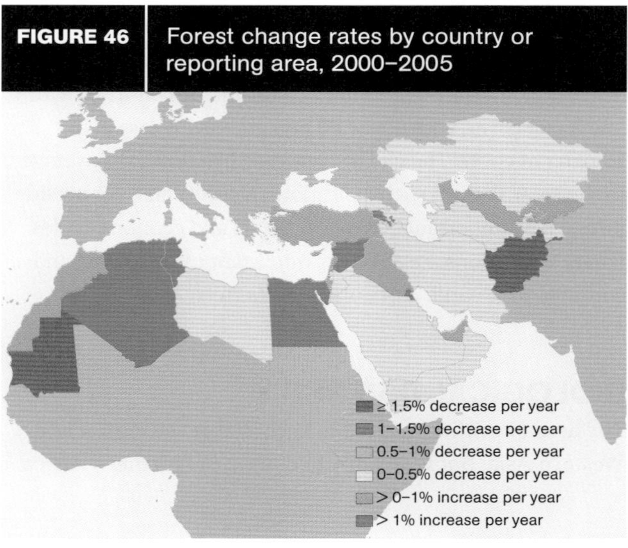


TABLE 26
**Extent and change of forest area**

| Subregion | Area (1 000 ha) | | | Annual change (1 000 ha) | | Annual change rate (%) | |
|---|---|---|---|---|---|---|---|
| | 1990 | 2000 | 2005 | 1990–2000 | 2000–2005 | 1990–2000 | 2000–2005 |
| Northern Africa | 84 790 | 79 526 | 76 805 | –526 | –544 | –0.64 | –0.69 |
| Central Asia | 15 880 | 15 973 | 16 017 | 9 | 9 | 0.06 | 0.06 |
| Western Asia | 27 295 | 27 546 | 27 570 | 25 | 5 | 0.09 | 0.02 |
| **Total Near East** | **127 966** | **123 045** | **120 393** | **–492** | **–530** | **–0.39** | **–0.43** |
| **World** | **4 077 291** | **3 988 610** | **3 952 025** | **–8 868** | **–7 317** | **–0.22** | **–0.18** |

TABLE 27
**Area of forest plantations**

| Subregion | Area (1 000 ha) | | | Annual change (1 000 ha) | |
|---|---|---|---|---|---|
| | 1990 | 2000 | 2005 | 1990–2000 | 2000–2005 |
| Northern Africa | 7 696 | 7 513 | 7 503 | −18 | −2 |
| Central Asia | 1 274 | 1 323 | 1 193 | 5 | −26 |
| Western Asia | 3 022 | 3 623 | 3 895 | 60 | 55 |
| **Total Near East** | **11 991** | **12 460** | **12 591** | **47** | **26** |
| **World** | **101 234** | **125 525** | **139 466** | **2 424** | **2 788** |

Algeria, Egypt, Morocco and Tunisia all experienced an increase in forest area in recent years as a result of increased forest plantations. However, the Sudan lost almost 12 percent of its forest area from 1990 to 2005. It remains the most forested country in the region, but this could change if actions are not taken to reverse the high rate of deforestation.

The total area of other wooded land is roughly equal to that of forest land area. However, data for other wooded land are incomplete, with several of the larger countries, including the Sudan, not having produced estimates for 2005.

Globally, forest plantations account for about 4 percent of total forest area. In the region, forest plantations account for about 10.5 percent of forest area (Table 27). Forest plantations play a particularly important role in several countries with low forest cover – for example in Kuwait, Oman and the United Arab Emirates, 100 percent of the forest area is made up of forest plantations.

In short, as might be expected in one of the world's driest regions, the Near East is dominated by countries with low forest cover, with about 80 percent having less than 10 percent cover. The global average is five times the average in the Near East. In such conditions, forests and trees outside forests play important ecological, social and economic roles. Forest plantations are also very important in the region and continue to expand, particularly in Western Asia.

## BIOLOGICAL DIVERSITY

The area of primary forests is fairly stable in Central and Western Asia, but is steadily declining in Northern Africa. As with forest area as a whole, the largest losses are taking place in the Sudan.

Although forest area designated primarily for conservation has increased slightly in the past five years, the area has been fairly stable since 1990 (Table 28). In contrast, this parameter has been increasing on a fairly regular basis in most other regions and in the world as a whole.

Other indicators of biological diversity include the number of tree species per country (Figure 47) and the number of species considered to be endangered or vulnerable. Based on the information available, there is no evidence that forest biological diversity is either substantially decreasing or increasing in the region.

**FIGURE 47** Number of native tree species

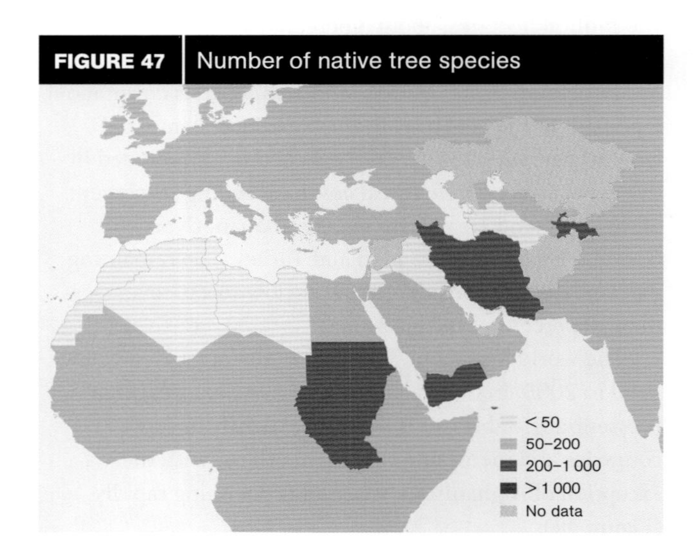


TABLE 28
**Area of forest designated primarily for conservation**

| Subregion | Area (1 000 ha) | | | Annual change (1 000 ha) | |
|---|---|---|---|---|---|
| | 1990 | 2000 | 2005 | 1990–2000 | 2000–2005 |
| Northern Africa | 9 773 | 9 051 | 8 687 | −72 | −73 |
| Central Asia | 856 | 1 095 | 1 663 | 24 | 114 |
| Western Asia | 888 | 1 031 | 1 098 | 14 | 13 |
| **Total Near East** | **11 516** | **11 176** | **11 448** | **−34** | **54** |
| **World** | **298 424** | **361 092** | **394 283** | **6 267** | **6 638** |

TABLE 29
**Forest fires in selected countries**

| | Average annual number of fires | Average annual area burned (ha) | Time period for available data |
|---|---|---|---|
| Algeria | 1 739 | 54 797 | 1991–2000 |
| Cyprus | 156 | 1 955 | 1995–2004 |
| Islamic Republic of Iran | – | 6 500 | 1998–2002 |
| Kazakhstan | – | 179 000 | 1998–2002 |
| Morocco | 315 | 3 340 | 1990–1999 |
| Turkey | 2 306 | 12 069 | 1988–2004 |

## FOREST HEALTH AND VITALITY

In Central Asia, fire accounts for about 50 percent of the area disturbed, whereas in Northern Africa and Western Asia, it accounts for about 10 percent or less. All disturbances are poorly reported for Northern Africa.

Fire and insect pests are the greatest threat to forest health in the region. However, the data are not highly reliable, as most countries do not maintain good records on forest disturbances.

Over the past few years, some severe dieback and decline phenomena have been affecting mainly junipers and cedars, which serve both productive and protective functions. The multiplicity of interrelated causes is being examined, and there is interest in establishing a regional information exchange network.

Examples of decline include *Juniperus procera* in the Asir highlands, Saudi Arabia; *Cedrus atlantica* in Algeria and Morocco, representing the world's genetic base for Atlantic cedars; *Cedrus libani* in Lebanon; *Juniperus phoenicea* in the Libyan Arab Jamahiriya; and *Juniperus polycarpus* in Kyrgyzstan and Oman.

In Lebanon, *Cedrus libani* was under serious threat from repeated defoliations caused by a new pest, the cedar web-spinning sawfly, *Cephalcia tannourinensis.* Fortunately, concerted efforts in management reduced the risk to local trees and gene stock and prevented transboundary spread.

Woody invasive species are also causing some concern in the region, such as mesquite (*Prosopis* spp.) in Oman, the Sudan and Yemen.

Near Eastern countries established an agreement to create the Near East Plant Protection Organization in 1993. The agreement has been ratified by eight countries (most recently the Syrian Arab Republic in July 2005), but two more ratifications are required for it to enter into force.

Forest fires also have a serious impact on forest health in a number of countries in the region. Data were available for six countries (Table 29) (FAO, 2006d).

In recent years, community-based fire management programmes have been developed that emphasize a broad approach to fire prevention and control. For example, an integrated fire management project with financial support from Italy is under way in the coastal areas of the Syrian Arab Republic. It aims to restore degraded coastal ecosystems through participatory approaches to fire management.

An effective response requires good information about forest resources; access to science and expertise to address the more serious threats; and a commitment to take effective action to counter the threats, including the commitment of financial and human resources.

## PRODUCTIVE FUNCTIONS OF FOREST RESOURCES

Some 36 percent of the forest area in the Near East is designated primarily for production, similar to the global average of 34 percent. However, there is a downward trend in forest area so designated, both in the region and in the world as a whole (Table 30).

TABLE 30
**Area of forest designated primarily for production**

| Subregion | Area (1 000 ha) | | | Annual change (1 000 ha) | |
|---|---|---|---|---|---|
| | 1990 | 2000 | 2005 | 1990–2000 | 2000–2005 |
| Northern Africa | 35 067 | 32 899 | 31 331 | –217 | –313 |
| Central Asia | 27 | 28 | 28 | n.s. | 0 |
| Western Asia | 9 539 | 9 563 | 9 513 | 2 | –10 |
| **Total Near East** | **44 633** | **42 490** | **40 872** | **–214** | **–323** |
| **World** | **1 324 549** | **1 281 612** | **1 256 266** | **–4 294** | **–5 069** |

**NOTE:** n.s. = not significant

TABLE 31
**Growing stock**

| Subregion | Growing stock (million m³) | | | Growing stock (m³/ha) | | |
|---|---|---|---|---|---|---|
| | 1990 | 2000 | 2005 | 1990 | 2000 | 2005 |
| Northern Africa | 1 436 | 1 409 | 1 390 | 17 | 18 | 18 |
| Central Asia | 1 004 | 1 041 | 1 061 | 63 | 65 | 66 |
| Western Asia | 1 959 | 2 069 | 2 111 | 72 | 75 | 77 |
| **Total Near East** | **4 399** | **4 520** | **4 562** | **34** | **37** | **38** |
| **World** | **445 252** | **439 000** | **434 219** | **109** | **110** | **110** |

Forest management for industrial wood production is limited to a few countries in the region, for example the Islamic Republic of Iran, the Sudan and Turkey. There is a history of wood production in Cyprus, but the emphasis in recent years has been to set aside forests for recreational purposes.

Growing stock in the region is increasing (Table 31). However, in the Near East it only represents about 1 percent of the global total, compared with 3 percent of the forest area. A relatively low growing stock per hectare is characteristic of arid and semi-arid forest ecosystems.

**FIGURE 48** Wood removals

Throughout the region, fuelwood is the major source of energy in rural households, where it is used for heating and cooking. About two-thirds of the wood in the Near East is used for fuel, compared with a global average of 40 percent (Figure 48). However, as fossil fuel prices rise, it can be anticipated that fuelwood use will increase in all parts of the world.

## PROTECTIVE FUNCTIONS OF FOREST RESOURCES

The trend in area of forest designated primarily for protective functions is positive (Table 32). This is an indication that governments recognize the importance of

**FIGURE 49** Designated primary functions of forests, 2005

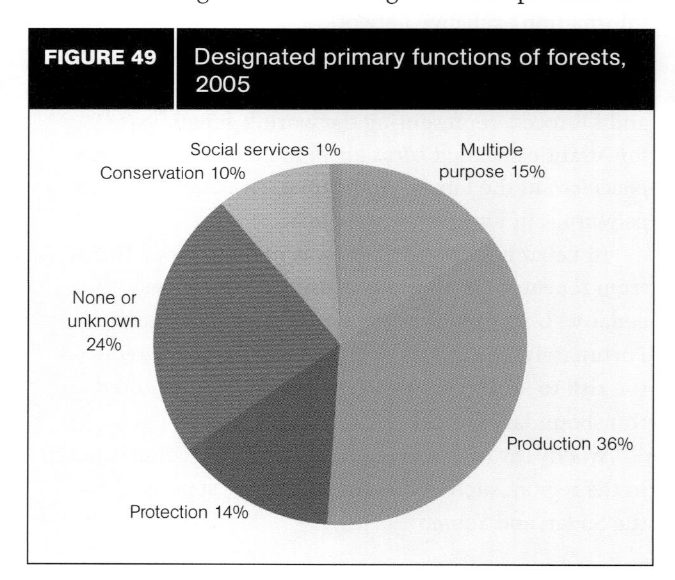


TABLE 32
**Area of forest designated primarily for protection**

| Subregion | Area (1 000 ha) | | | Annual change (1 000 ha) | |
|---|---|---|---|---|---|
| | 1990 | 2000 | 2005 | 1990–2000 | 2000–2005 |
| Northern Africa | 3 645 | 3 819 | 3 861 | 17 | 8 |
| Central Asia | 10 328 | 10 958 | 10 962 | 63 | 1 |
| Western Asia | 1 751 | 1 974 | 2 085 | 22 | 22 |
| **Total Near East** | **15 724** | **16 752** | **16 908** | **103** | **31** |
| **World** | **296 598** | **335 541** | **347 217** | **3 894** | **2 335** |

the protective functions of forests and trees, for example in combating desertification. The area designated for protective functions in 2005 was about 14 percent of total forest area, compared with a global average of about 8 percent. However, not all countries use this designation, and some protective functions may be included under "multiple purpose" (Figure 49).

About 35 percent of forest plantations were designated primarily for protection, compared with a global average of about 20 percent.

## SOCIO-ECONOMIC FUNCTIONS

The value added by the forest sector in one year in the Near East is about US$5 billion.

The value added of the forest sector in the Near East was somewhat volatile during the 1990s, peaking in 1995 (Figure 50). The percentage contribution of the forest sector to the overall regional economy is declining steadily, owing in large part to overall economic growth in the region: other key sectors are growing, particularly oil, while the forest sector is relatively stable.

The value of imported forest products is almost five times the value of exports. Forest products account for a declining percentage of the total value of all goods traded, both in the region and globally. The value of forest products traded has risen substantially, but the value of goods traded in other sectors has risen even more dramatically.

The highest-value forest products imported into the region are primary paper products and primary wood products, such as plywood, lumber and particle board, followed by secondary products such as furniture and other products manufactured from wood (Figure 51). This is a positive sign, because it indicates that a significant share of the manufacturing of secondary products is taking place within the region, thus creating income and employment.

**FIGURE 51** Trends in net trade of forest products by subsector

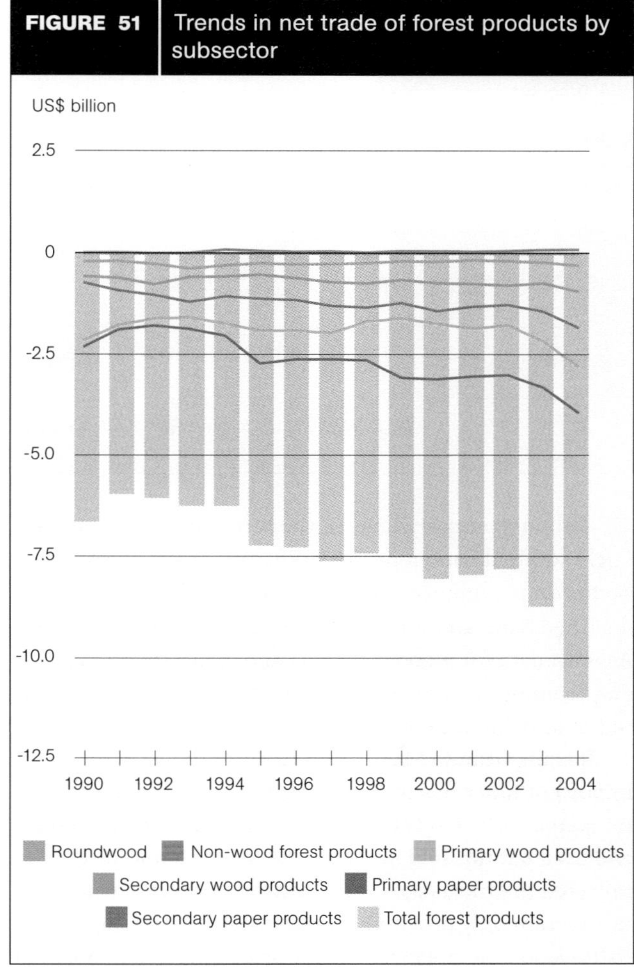


**NOTE:** A positive value indicates net export. A negative value indicates net import.

**FIGURE 50** Trends in value added in the forest sector, 1990–2000

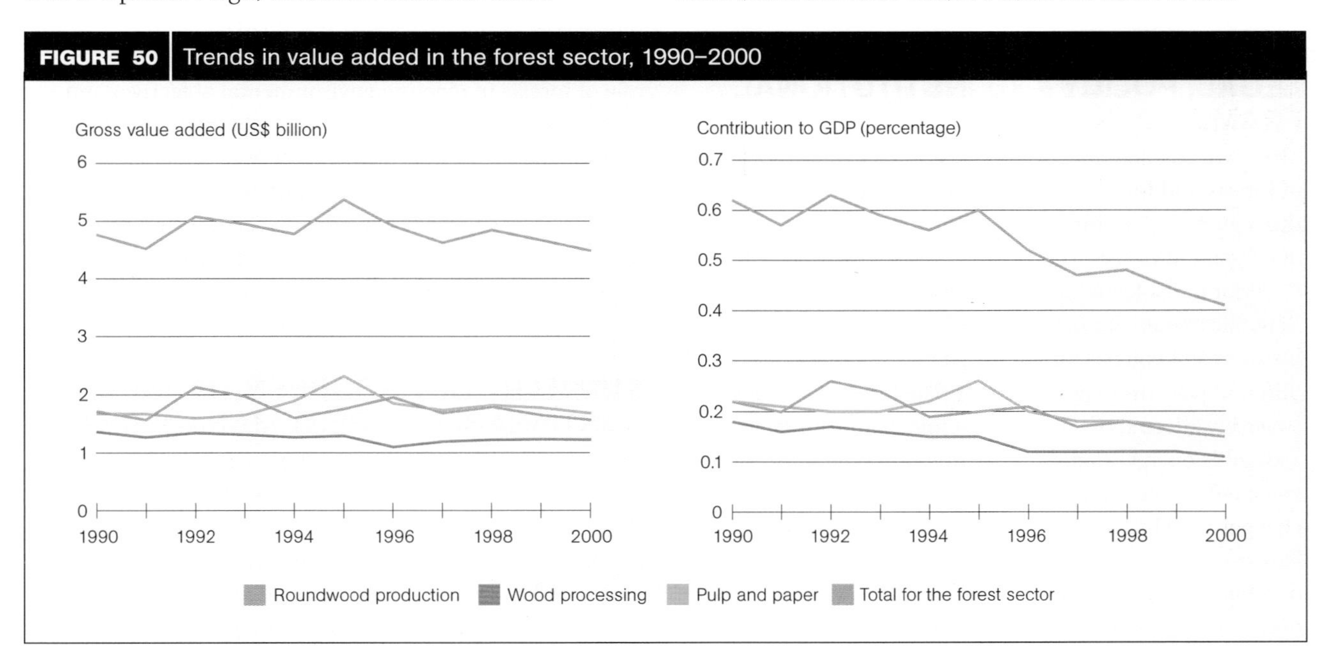

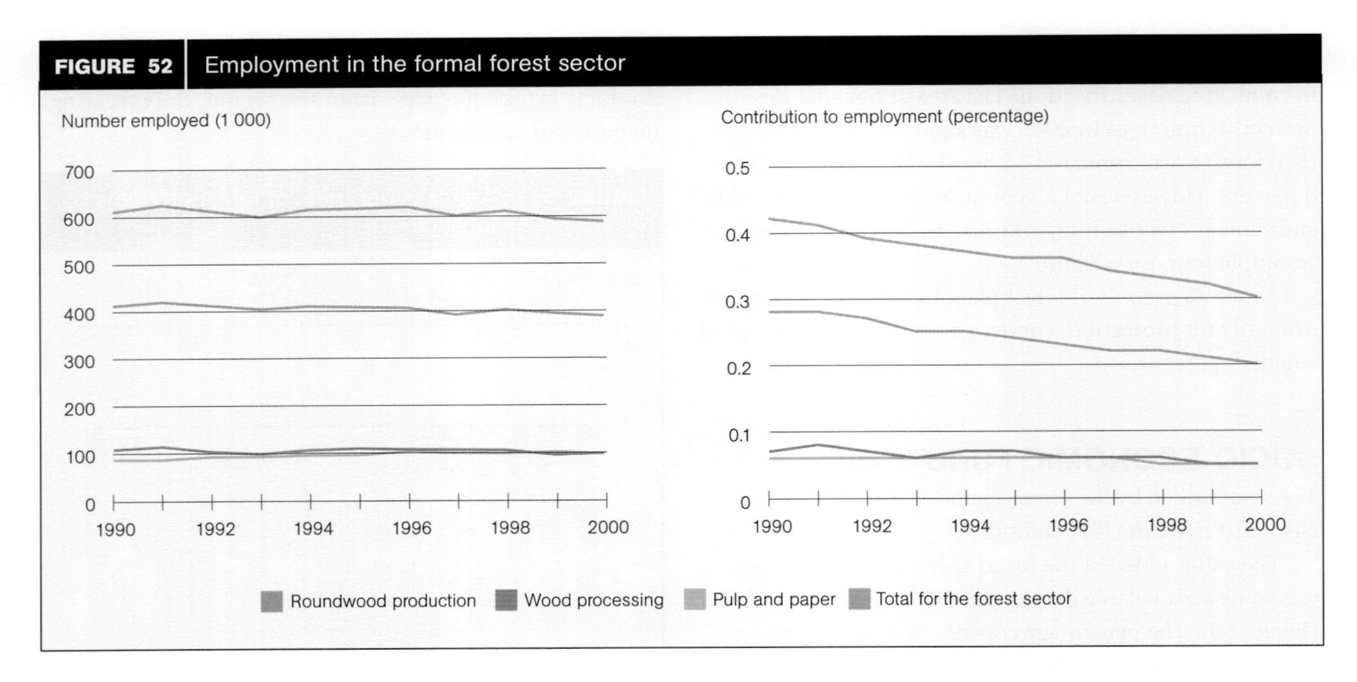


Although employment in the forest sector remained fairly stable throughout the 1990s, the share of forest-sector employment in total employment in the region declined from about 0.4 to 0.3 percent (Figure 52). As with data for wood removals and value added the employment data indicate that forestry is a relatively flat industry, while other key sectors are growing.

It is important to recall that many of the most important functions of forests are not valued in the marketplace. The NWFPs and fuelwood that are gathered and used but not sold in the market are also not fully reflected in official economic statistics. Thus the data in this section only provide a partial basis for assessing the socio-economic importance of forests. This is a dilemma of forestry in the Near East and other regions.

## LEGAL, POLICY AND INSTITUTIONAL FRAMEWORK

There are considerable differences in the evolution of forests and forestry among different countries in the region, depending on the individual histories and development trajectories (FAO, 2006h).

Prior to the break-up of the Union of Soviet Socialist Republics, countries in Central Asia shared common histories and policies but, in the past 15 years, they have had different patterns of development. The forest sector has been adversely affected because of a decline in an affordable and accessible energy supply, a reduction in timber availability and a reduction in human and financial resources. These changes have had a generally negative impact on forests. However, some economies have subsequently begun to develop rapidly as countries have learned to adapt to more open economies and political processes.

Countries in the Near East are heavily influenced by the external political and economic environment. Because of its dominant role in global energy supply, the Near East is more affected by external global forces than are most other regions of the world.

A number of countries have demonstrated political commitment to forests over the past 15 years. Among those that have enacted new forest policies or laws are Morocco, Saudi Arabia, the Sudan, the Syrian Arab Republic, Tunisia, Turkey and Uzbekistan (FAO, 2006e). Countries with forestry educational institutions include Algeria, Cyprus, Egypt, the Islamic Republic of Iran, Iraq, Morocco, Saudi Arabia, the Sudan, the Syrian Arab Republic and Turkey.

Responsibility for forest management has been transferred to the environment ministry in many countries, reflecting a growing recognition of the potential role of forests in meeting environmental objectives and perhaps a declining role for their productive functions. A problem in a number of countries is a lack of clarity regarding the responsibilities of different institutions for forest and rangeland management. Competition among ministries and agencies reduces the effectiveness of forest management in some countries.

## SUMMARY OF PROGRESS TOWARDS SUSTAINABLE FOREST MANAGEMENT

Progress is being made in a number of areas. In many countries in the region, forest cover is stable and deforestation is not a big problem. Leaders throughout the region have recognized the importance of forests, and most countries have taken steps to expand and protect forests through laws, policies and programmes.

It is not surprising that the countries having the most difficulty managing their forests and controlling deforestation are those experiencing conflict, including Afghanistan, Iraq and the Sudan.

A key limiting factor for countries striving to improve the management of their forests is the absence of adequate resources. Most forest resources are publicly owned, but either public resources are increasingly scarce or the allocated share of the public budget is inadequate. The forest sector must do a better job of making the benefits of forests known to political decision-makers, as well as promoting sustainable private-sector investment in forests.

Some countries in other regions have been successful in using incentives for good forest management, as well as experimenting with payments for environmental services. The potential of this approach in the Near East has yet to be fully explored.

Despite the problems and limitations faced by countries in the Near East, experience has shown that progress can be and is being made through effective strategies for mobilizing knowledge and resources.

**FIGURE 53** Subregional breakdown used in this report

For purposes of this report, North America includes Canada, Mexico and the United States of America (excluding United States territories in the Caribbean).

# North America

## EXTENT OF FOREST RESOURCES

Forests cover 33 percent of North America's land area (Figure 54) and account for 17 percent of global forest area. The world lost about 3 percent of its forest area from 1990 to 2005; but in North America, total forest area remained virtually constant (Table 33 and Figure 55). Canada reported no change in forest area from 1990 to 2005; Mexico reported a decrease of 0.52 percent per year from 1990 to 2000, which slowed to a decrease of 0.40 percent per year from 2000 to 2005. The United States of America reported an annual increase in forest area of 0.12 percent in the 1990s and 0.05 percent from 2000 to 2005.

Forest plantations, which represent about 4 percent of total forest area globally, account for 5.6 percent of forest area in the United States of America and 1.6 percent in Mexico (Table 34). Canada was not able to report on this parameter in the context of FRA 2000 or FRA 2005.

The extent of forests in North America is fairly stable, which is especially significant when compared with the world as a whole. The continuing loss of forest cover in Mexico remains a concern, although the percentage rate is less than in many other countries. For example, Mexico's neighbour Guatemala is losing forest area at a rate more than three times that of Mexico.

Forest plantations make up an increasing proportion of total forest area in some parts of the world. For example, China alone has four times the area of planted forests of the United States of America (FAO, 2006i). The absence

TABLE 33
**Extent and change of forest area**

| Subregion | Area (1 000 ha) | | | Annual change (1 000 ha) | | Annual change rate (%) | |
|---|---|---|---|---|---|---|---|
| | 1990 | 2000 | 2005 | 1990–2000 | 2000–2005 | 1990–2000 | 2000–2005 |
| Canada | 310 134 | 310 134 | 310 134 | 0 | 0 | 0 | 0 |
| Mexico | 69 016 | 65 540 | 64 238 | –348 | –260 | –0.52 | –0.40 |
| United States of America | 298 648 | 302 294 | 303 089 | 365 | 159 | 0.12 | 0.05 |
| **Total North America** | **677 798** | **677 968** | **677 461** | **17** | **–101** | **0** | **–0.01** |
| **World** | **4 077 291** | **3 988 610** | **3 952 025** | **–8 868** | **–7 317** | **–0.22** | **–0.18** |

**FIGURE 54** Extent of forest resources

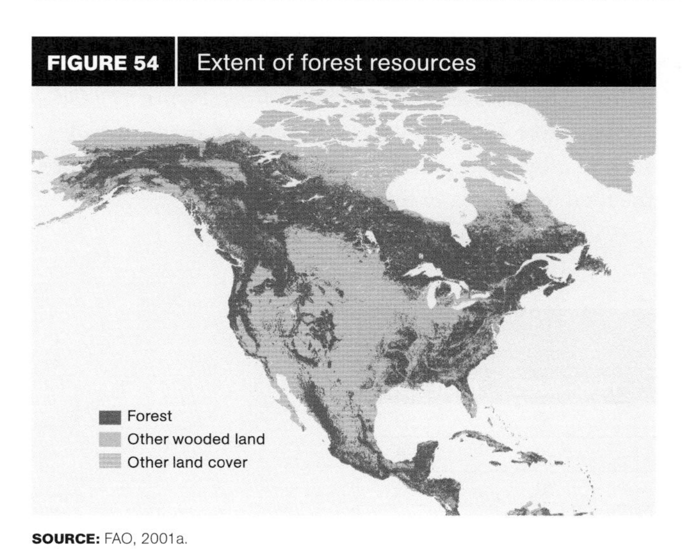


**SOURCE:** FAO, 2001a.

**FIGURE 55** Forest change rates by country or reporting area, 2000–2005

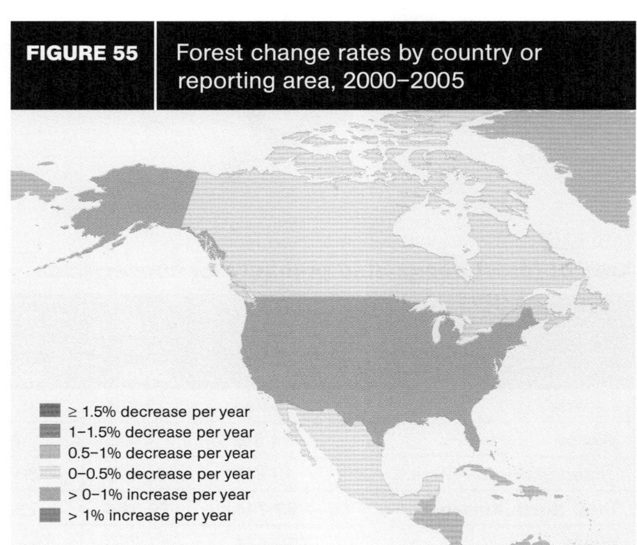

TABLE 34
**Area of forest plantations**

| Subregion | Area (1 000 ha) | | | Annual change (1 000 ha) | |
|---|---|---|---|---|---|
| | 1990 | 2000 | 2005 | 1990–2000 | 2000–2005 |
| Canada | – | – | – | – | – |
| Mexico | – | 1 058 | 1 058 | – | 0 |
| United States of America | 10 305 | 16 274 | 17 061 | 597 | 157 |
| **Total North America** | **10 305** | **17 332** | **18 119** | **–** | **157** |
| **World** | **101 234** | **125 525** | **139 466** | **2 424** | **2 788** |

**NOTE:** The North America total figures only refer to the country (ies) that have reported on this variable.

of information about forest plantations in Canada makes it difficult to draw conclusions about this particular parameter for North America as a whole.

## BIOLOGICAL DIVERSITY

Primary forests account for 45 percent of the forests in the region, with more than half the total found in Canada. These forests are stable in Canada, declining at an annual rate of 1.1 percent in Mexico and declining slightly in the United States of America. In contrast, at the global level, primary forests are declining at an annual rate of about 0.6 percent.

Some 12 percent of North America's forest area is designated for conservation, compared with a global average of 11 percent. Forest area designated for conservation increased in the United States of America at an annual rate of 3.7 percent from 2000 to 2005, with Canada reporting no change and Mexico reporting an annual decrease of 0.2 percent (Table 35). At the regional level, the annual increase of 2.7 percent exceeds the global average increase of 1.8 percent.

Other indicators of biological diversity include the number of tree species per country (Figure 56) and the number of tree species considered to be endangered or vulnerable. In general, the diversity of all species, including tree species, increases with proximity to the equator.

It is not possible to make sweeping conclusions about trends based on this information; there is little evidence that forest biological diversity in the region is either substantially decreasing or increasing.

Within the region, Mexico is the area of greatest concern, as it is experiencing significant losses in its primary forests. At the global level, tropical and dryland forest ecosystems are under the greatest pressure, and Mexico has significant forest area in both of these categories.

**FIGURE 56** Number of native tree species

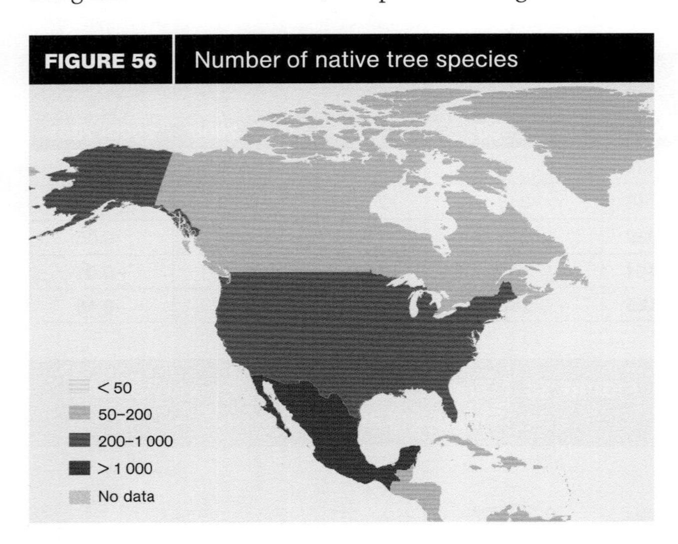


## FOREST HEALTH AND VITALITY

In the region as a whole, insects account for the highest proportion of disturbances in terms of forest area, followed by diseases and then forest fires (Figure 57). On average, over 40 million hectares of forest are disturbed each year by insects, disease or fire, or about 6 percent of the total forest area in the region.

TABLE 35
**Area of forest designated primarily for conservation**

| Subregion | Area (1 000 ha) | | | Annual change (1 000 ha) | |
|---|---|---|---|---|---|
| | 1990 | 2000 | 2005 | 1990–2000 | 2000–2005 |
| Canada | 15 284 | 15 284 | 15 284 | 0 | 0 |
| Mexico | 4 513 | 4 425 | 4 381 | –9 | –9 |
| United States of America | 49 948 | 50 675 | 60 076 | 73 | 1 880 |
| **Total North America** | **69 745** | **70 384** | **79 741** | **64** | **1 871** |
| **World** | **298 424** | **361 092** | **394 283** | **6 267** | **6 638** |

**FIGURE 57** Forest disturbances, 1998–2002

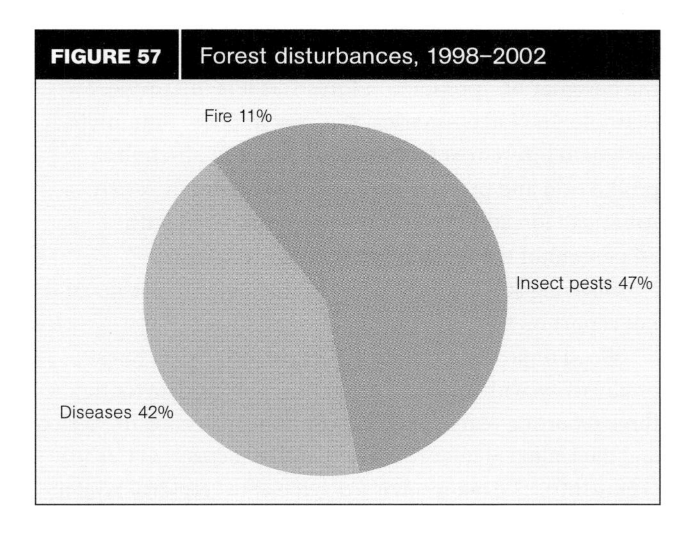


North American forests have long been subjected to pressure by both indigenous and exotic pest outbreaks, affecting trade and ecosystem functions and increasing fire and safety hazards. Introduced insect pests include the Asian longhorn beetle, *Anoplophora glabripennis*, and the emerald ash borer, *Agrilus planipennis*. Examples of diseases include sudden oak death, *Phytophora ramorum*, and Eucalyptus rust, *Puccinia psidii*, a recent arrival in Hawaii, United States of America. Naturally occurring forest insect pests include the mountain pine beetle, *Dendroctonus ponderosae*, and the southern pine beetle, *Dendroctonus frontalis* – the most destructive insect pest of pine forests in the southern United States of America and in parts of Mexico (Payne, 1980).

The detection of the emerald ash borer in southeastern Michigan, United States of America, in 2002 has focused attention on invasive species entering the region, becoming established and remaining undetected for some time. This exotic beetle probably arrived in the United States of America on solid wood packing material from its native Asia. It subsequently became established in the central United States of America and Canada, where it has killed more than 20 million ash trees and contributed to regulatory changes and enforced quarantine. The cost to municipalities, property owners, nursery operators and forest products industries has run into tens of millions of dollars. Measures have been taken to store viable seed source so that the ash tree population can be re-established in the event that the borer cannot be contained.

It is not only exotic invasive pests that cause significant management challenges. At times, managing outbreaks of indigenous pests is as much of a challenge as managing some exotics. The extent and intensity of such outbreaks can be affected by other types of disturbances such as fire, extreme weather events or human activity. An example is the recent outbreak of the naturally occurring mountain pine beetle in Canada. In 2005 it was estimated that, since 1997, this outbreak had affected from 7 to 8.5 million hectares of forest. It was predicted that, by the end of 2006, approximately 40 percent of the susceptible pine would have been killed or harvested. In previous outbreaks, mountain pine beetles have killed as many as 80 million trees distributed over 450 000 ha, making them the second most important natural disturbance agent after fire in these forests. The Government of British Columbia has dramatically increased logging in an attempt to slow the spread of the beetle by removing recently infested trees and to recover value from trees already killed. To handle increased harvest allowances, the forest industry has increased its capacity to process the wood. The environmental and social implications of this epidemic are under surveillance.

Forest fires have a major impact on forest health in all three North American countries, and in recent years fire management has been one of the major preoccupations of forest leaders. All three countries have observed long-term increases in the severity of fires and in losses from catastrophic fires (Figure 58). The three countries regularly share information and resources in an effort to prevent and manage forest fires.

Dramatic swings from year to year are an indication of the effect of changes in climate: severe fire seasons are often followed by relatively mild seasons. This variability poses severe management challenges for agencies planning budgets and human resources without knowing the severity of the next fire season. Programmes to prevent unwanted fires and to manage the beneficial aspects of fire are becoming more sophisticated and more costly. Fire plays an important role in many forest ecosystems in the

**FIGURE 58** Wildland fires

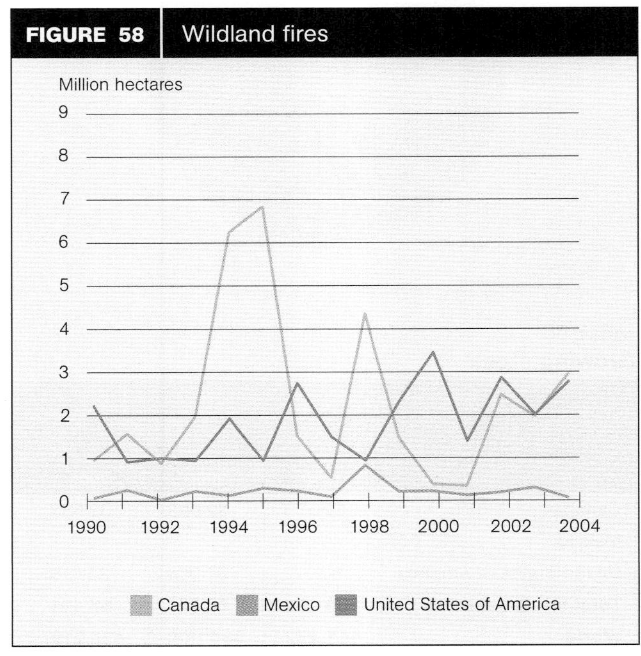


**SOURCE:** FAO, 2006d.

region, particularly boreal forests, and prescribed fire is an important, although risky, forest management tool.

North America has several regional mechanisms for promoting cooperation on forest health issues. The working groups for forest insects and disease and for forest fire management are the longest established groups of the North American Forest Commission (NAFC), each having been established over 40 years ago. Recently, a new working group on invasive species has been added. In addition, the North American Plant Protection Organization, recognized under the IPPC, offers mechanisms for regional coordination on phytosanitary matters, including those for reporting on pests and activating alerts, as well as providing fact sheets.

## PRODUCTIVE FUNCTIONS OF FOREST RESOURCES

The production of wood and non-wood products is very important in the region, with a major impact on the social and economic dimensions of sustainable development. About 6 percent of forest land is designated primarily for production, compared with 32 percent at the global level. However, this is a misleading statistic, as it is not a common practice of North American countries to use the designation "production forest" as it is in Europe, for example. In North America, it is much more common to designate forest area for "multiple purpose", including production and protection. Some 79 percent of North America's forests are so designated, compared with 34 percent of the world's forests (Figure 59).

Growing stock is increasing (Table 36), but Mexico did not report on this variable.

Wood removals provide another perspective on forest productivity, with all three countries reporting data for the three reporting years. Removals continue to decline in Mexico and the United States of America, and to increase in Canada (Figure 60). The net result is a decline at the regional level in the 1990s and a slight increase from 2000 to 2005.

There was a sharp decline in the 1990s in the use of fuelwood in Canada and the United States of America, although this trend has levelled off since 2000. It has continued to increase in Mexico since 1990. Moreover, it is likely to increase in all countries in the future if the price of fossil fuels continues to rise, providing an incentive to use wood and other renewable energy sources.

**FIGURE 59** Designated primary functions of forests, 2005

None or unknown 3%
Production 6%
Conservation 12%
Multiple purpose 79%
Social services 0%
Protection 0%

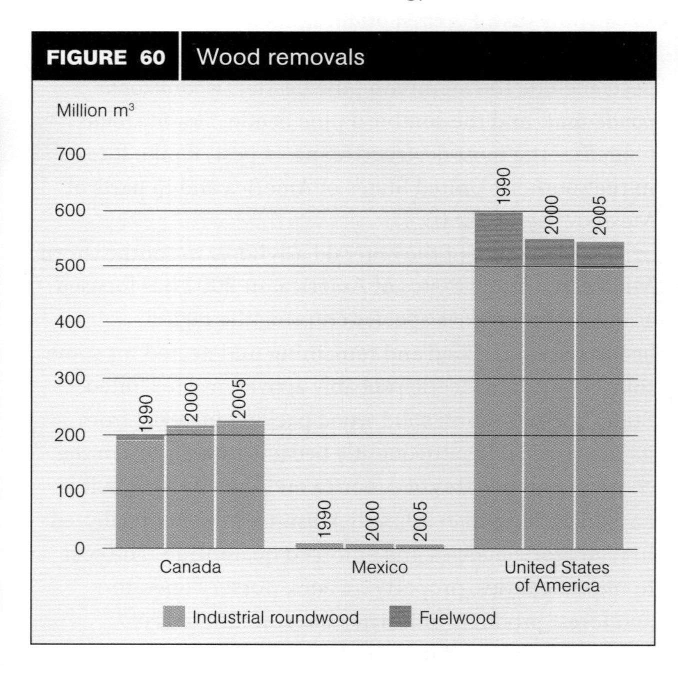


**FIGURE 60** Wood removals

TABLE 36
**Growing stock**

| Subregion | Growing stock | | | | | |
|---|---|---|---|---|---|---|
| | *(million m³)* | | | *(m³/ha)* | | |
| | 1990 | 2000 | 2005 | 1990 | 2000 | 2005 |
| Canada | 32 983 | 32 983 | 32 983 | 106 | 106 | 106 |
| Mexico | – | – | – | – | – | – |
| United States of America | 32 172 | 34 068 | 35 118 | 108 | 113 | 116 |
| **Total North America** | **65 155** | **67 051** | **68 101** | **107** | **109** | **111** |
| **World** | **445 252** | **439 000** | **434 219** | **109** | **110** | **110** |

**NOTE:** The North America total figures only refer to the country (ies) that have reported on this variable.

North America accounts for 40 percent of the world's wood removals and only 17 percent of the world's forest area, suggesting that the region's forests are relatively productive and the use of forests for commercial purposes is relatively well advanced.

Only 7 percent of the wood removed in North America is used for fuel, compared with a global average of 40 percent. In contrast, in Africa, fuelwood accounts for almost 90 percent of wood removals.

The information available on NWFPs at the regional level is not sufficient to draw conclusions or to identify trends. However, there is evidence that the use of forests for a variety of such products is increasing in many parts of the region.

## PROTECTIVE FUNCTIONS OF FOREST RESOURCES

As mentioned in the previous section, in North America, a majority of the forest area is designated as multiple purpose, including production and protection (Figure 59). None of the countries uses the category "forest designated primarily for protection", and only Mexico reported any protective forest plantations.

Mexico is one of the world leaders in an emerging area of innovative public policy – payment for environmental services. As of 2005, over 500 000 ha of forest in Mexico were covered by schemes to pay forest owners for the benefits of good forest management, providing clean water and mitigating climate change.

Although the protective functions of forests are well known, and these values seem to be increasingly recognized and discussed in the popular media as well as in government and academic circles, there is a shortage of macrolevel information to suggest whether trends are up or down at the regional level. This is an area requiring more applied research.

## SOCIO-ECONOMIC FUNCTIONS

The value added by forest products to national economies in the 1990s was generally upward, peaking in 1995 when prices for wood products were strong (Figure 61). However, the contribution of forests to GDP declined during the same period, mainly owing to the strengthening of other economic sectors.

Forest products are especially important in Canada's economy, contributing 12 percent of total export value in 2004 (compared with 3.4 percent in the United States of America and 1.3 percent in Mexico), although this is down from 15.5 percent as recently as 1990.

The most startling trend in trade of forest products is the dramatic increase in imports into the United States of America and, to a lesser extent, into Mexico and Canada. Canadian and Mexican exports continue to increase, while United States exports grew in the 1990s and have declined slightly since 2000. For the region as a whole, which was a strong net exporter in the 1990s, forest products imports are increasing far more rapidly than exports, and the region is now a net importer (Figure 62).

The United States of America, which was a net exporter until 15 years ago, is increasingly a net importer. The main reason for the trend is not a decline in exports, which are fairly stable, but the increase in imports, which grew from US$22 billion in 1990 to US$62 billion in 2004 (Figure 63). Today the value of forest products imports is more than double the value of exports – a trend driven primarily by secondary forest products. The trend

**FIGURE 61** Trends in value added in the forest sector, 1990–2000

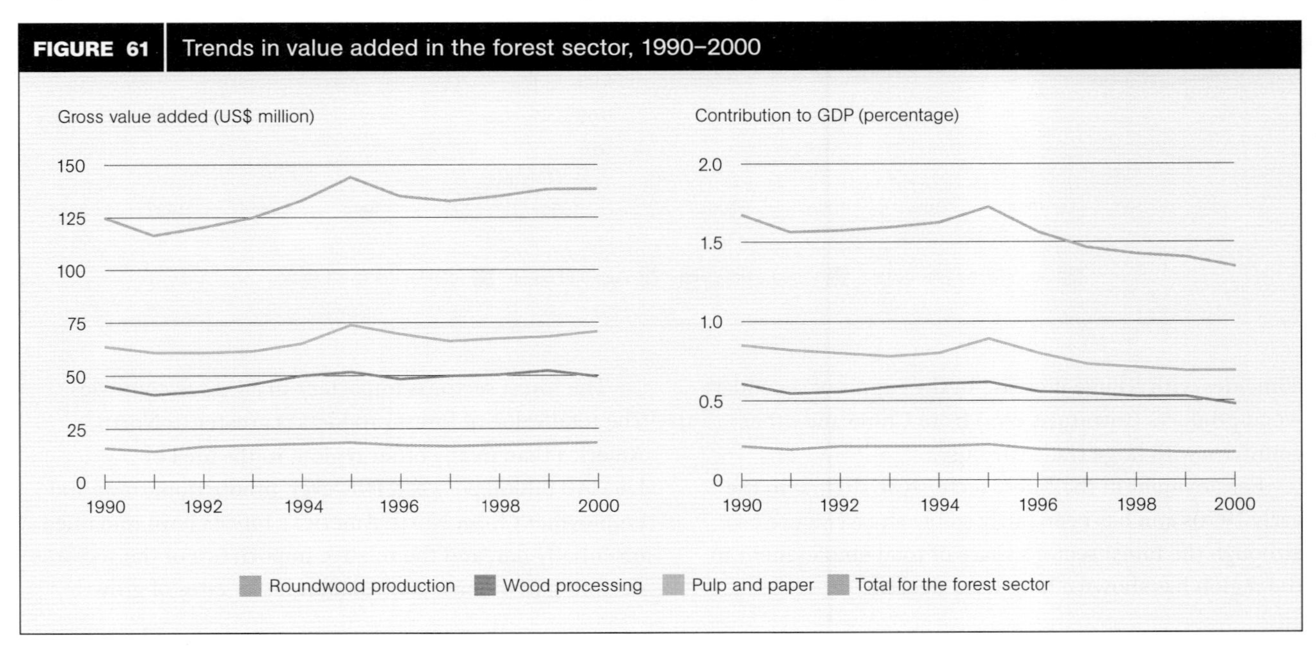

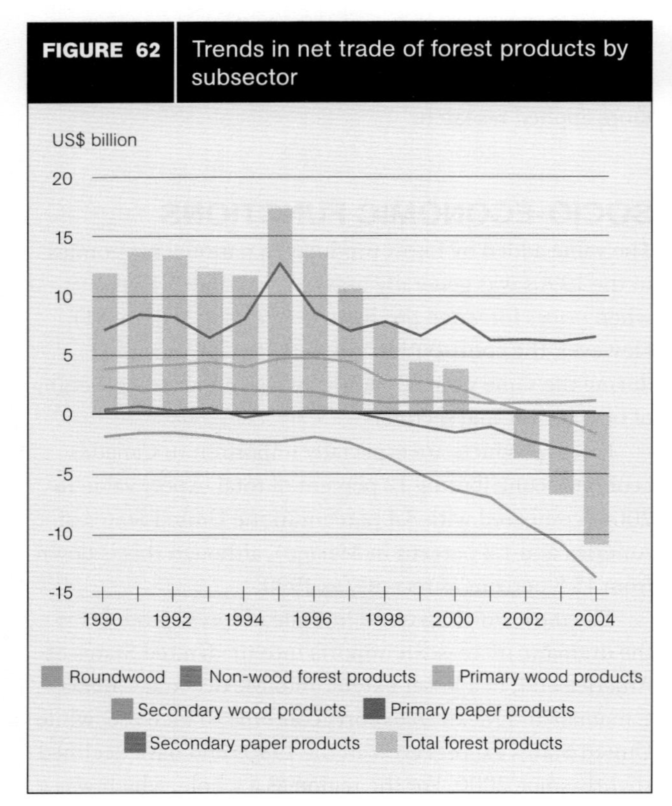


**FIGURE 62** Trends in net trade of forest products by subsector

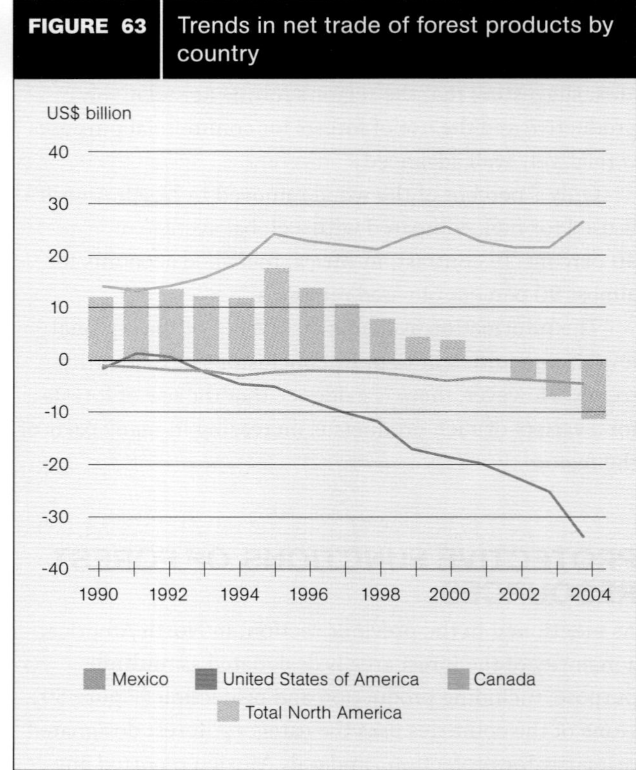


**FIGURE 63** Trends in net trade of forest products by country

**NOTE:** A positive value indicates net export. A negative value indicates net import.

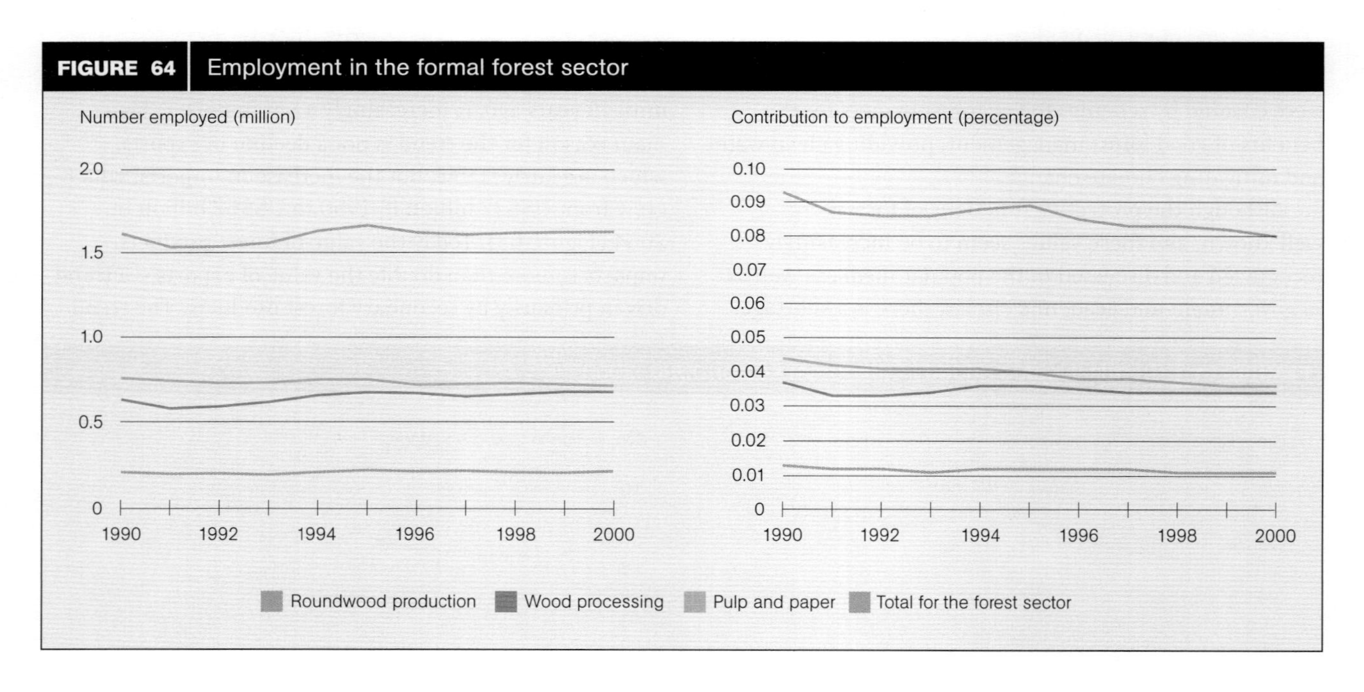


**FIGURE 64** Employment in the formal forest sector

coincides with a huge increase in the export of secondary wood products (furniture, etc.) from China and other countries with large trade surpluses.

Employment in the forest sector rose slightly in the early 1990s and has been fairly stable since (Figure 64), although the forest sector's share of total employment in the region has shown a long-term decline.

The forest sector is important in all three countries. The total value of forest products is greater in North America than in any other region, in the area of US$140 billion per year. However, production, trade and employment from North America's forests have remained essentially flat, and the relative importance of the sector is declining as other sectors exhibit faster rates of growth.

Wood removals are declining in Mexico and the United States of America, while they continue to increase in Canada. This trend is reflected in economic data, with modest growth in several economic indicators in Canada and a slight decline in the other two.

The data in this section do not reflect recreational uses of forests. FAO does not systematically collect data on this dimension, but the economic impact is significant. For example, the Forest Service of the United States Department of Agriculture (USDA) estimated that the outdoor recreation programme in national forests contributed US$11.2 billion to the United States national economy in 2002 (USDA, 2006).

## LEGAL, POLICY AND INSTITUTIONAL FRAMEWORK

For a country to achieve sustainable forest management, it must have a supportive legal, policy and institutional framework. In this respect, the North American region clearly has a solid foundation. All three countries have progressive policies that promote a mixture of private enterprise and public controls, decentralization within a stable national framework, clearly designated access rights to forest resources, and functioning forest research and educational institutions.

In Mexico, the only country in the region currently experiencing deforestation at the national level, the most serious problems do not appear to be related to institutional failures. On the contrary, Mexico has recently made changes that are already showing major benefits, such as the new forest law and the establishment of the National Forest Commission. The country is struggling to develop its economy, and it is sometimes easy to forget that Canada and the United States of America also experienced significant deforestation during periods of rapid population and economic growth. For example, forest area in the United States of America today is estimated at 28 percent less than it was at the time of European settlement. Most of the loss occurred during the period of rapid westward expansion from 1850 to 1900.

The institutional framework for forest management is significantly different in each of the three countries, resulting in large part from diverse ownership patterns. In Canada, 92 percent of the forests are publicly owned, almost all under the responsibility of the provinces. In Mexico, 59 percent of the forests are public, while, in the United States of America, public forests account for 42 percent of the total. What sets Mexico apart from its northern neighbours is the nature of its public forests, which are mainly in the form of *ejidos*. These are forests that are managed and whose benefits are shared by local communities. Mexico is one of the world's most advanced models of community forest management.

In sum, the underlying framework reflects a strong political commitment to achieve sustainable forest management in all three countries.

## SUMMARY OF PROGRESS TOWARDS SUSTAINABLE FOREST MANAGEMENT

With respect to most of the thematic elements, North America is making more progress than most other regions, especially those with a high proportion of developing countries or countries with economies in transition.

All three countries of the region are particularly concerned about forest health, and they have undertaken collaboration to address transboundary issues in this area. Working groups under the North American Forest Commission address fire, invasive species, and forest insects and disease at the regional level.

There is an obvious inverse correlation between economic development and deforestation. It is not surprising that Mexico, with the lowest per capita GDP in North America, is the only country in the region struggling with deforestation. At the other economic extreme, the United States of America is facing the problem of a declining forest industry, as indicated by weak employment and a rapidly increasing trade deficit in wood and paper products.

Sustainable forest management would seem to be an attainable goal in North America.

# The global view

## EXTENT OF FOREST RESOURCES

The world has just under 4 billion hectares of forest, covering about 30 percent of the world's land area. Forests are unevenly distributed around the world: of 229 countries or other reporting areas in FRA 2005, 43 have forest area exceeding 50 percent of their total land area, while 64 have forest area of less than 10 percent (Figure 65). Five countries (the Russian Federation, Brazil, Canada, the United States of America and China) together account for more than half the total forest area.

Deforestation continues at an alarming rate of about 13 million hectares a year. At the same time, forest planting and natural expansion of forests have significantly reduced the net loss of forest area.

Over the 15 years from 1990 to 2005, the world lost 3 percent of its total forest area, an average decrease of some 0.2 percent per year (Figure 66). From 2000 to 2005, the net rate of loss declined slightly – a positive development. In the same period, 57 countries reported an increase in forest area, and 83 reported a decrease (including 36 with a decrease greater than 1 percent per year). However, the net forest loss remains 7.3 million hectares per year or 20 000 ha per day.

Carbon stocks in forest biomass decreased by about 5.5 percent at the global level from 1990 to 2005. Regional trends generally follow the trends in forest area and growing stock: carbon stocks are increasing in Europe and North America and decreasing in tropical regions.

## BIOLOGICAL DIVERSITY

This theme includes so many interrelated variables that it is difficult to identify trends. Perhaps the most positive one is that many countries are increasing the forest area designated for conservation. From 1990 to 2005, the area so designated increased by 32 percent, a total increase of 96 million hectares, with increases in all regions. Globally, more than 11 percent of total forest area has been designated primarily for conservation of biological diversity (Figure 67).

Globally, 36 percent of forests are categorized as primary forests (forests of native species in which there are no clearly visible indications of human activity and ecological processes are not significantly disturbed). The leader is Latin America and the Caribbean (75 percent), followed by North America (45 percent).

At the global level, an estimated 6 million hectares of primary forest are lost or modified each year. Nine of the ten countries that account for more than 80 percent of the world's primary forest area lost at least 1 percent of this

**FIGURE 65** Forest area, 2005

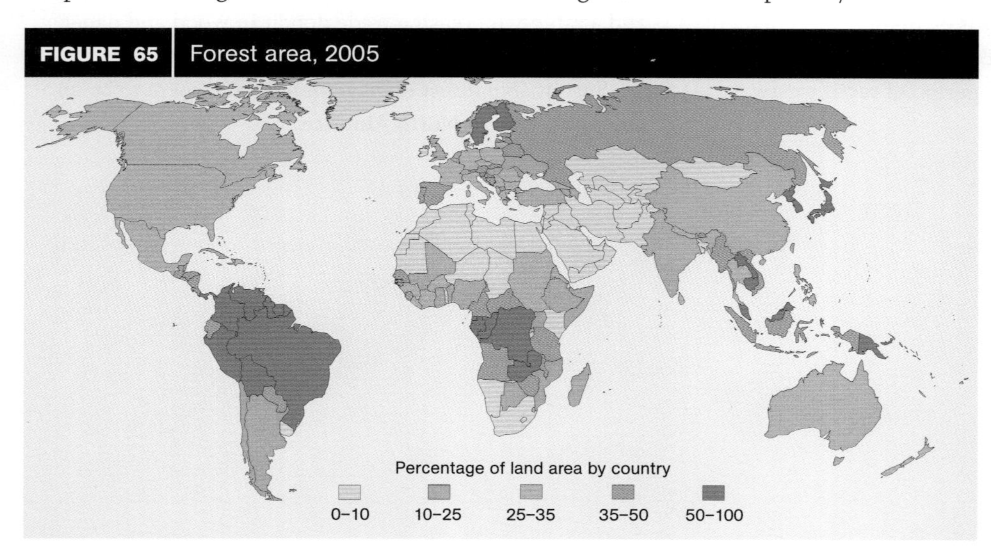

**FIGURE 66** Annual net change in forest area, 2000–2005

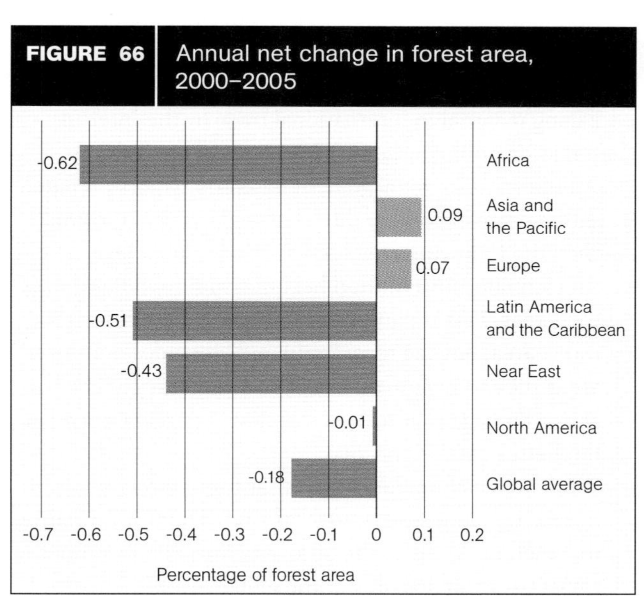


**FIGURE 67** Percentage of forest area designated for conservation, 2005

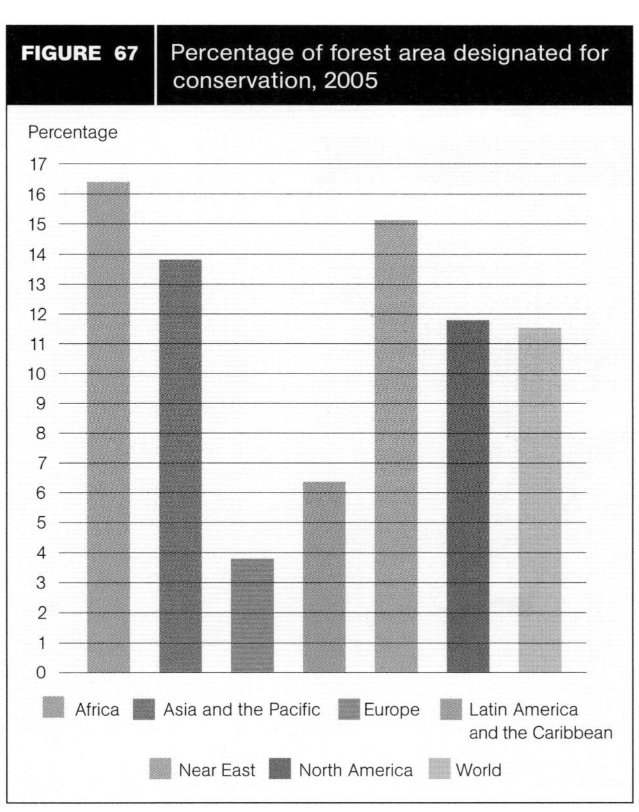


**FIGURE 68** Vulnerable and endangered tree species

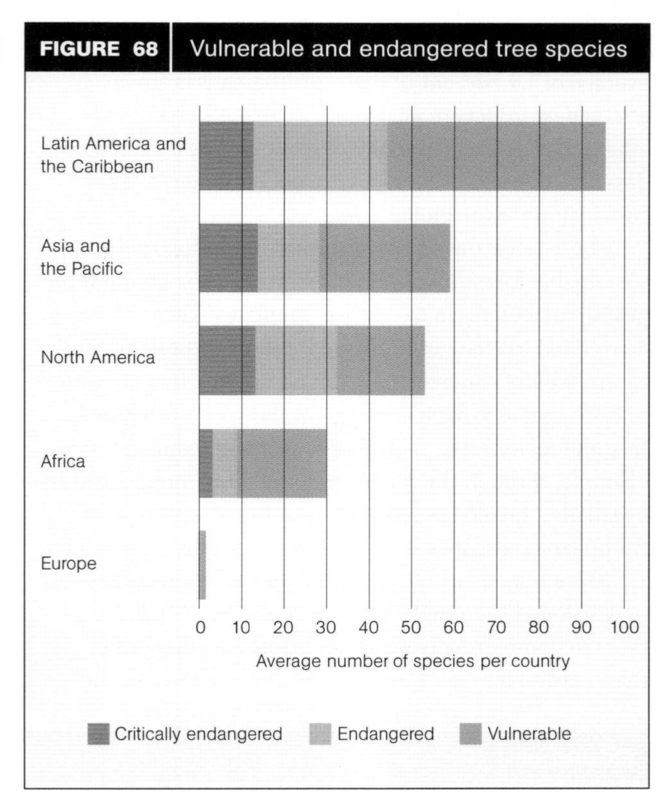


area from 2000 to 2005, led by Indonesia (13 percent loss in just five years), Mexico (6 percent), Papua New Guinea (5 percent) and Brazil (4 percent).

Another indicator of biological diversity is the number of threatened or endangered species (Figure 68). A majority of vulnerable and endangered tree species are found in tropical countries. Recently established baseline data will facilitate the identification of trends in the future.

In summary, there is good and bad news. The increase in forest designated for conservation is a positive trend, indicating political will in many countries to conserve biological diversity. However, the continuing decline in primary forests in most tropical countries is a matter of serious concern. While there are insufficient trend data to conclude that forest biological diversity is declining at a specific rate at the global level, there is nonetheless a clear downward trend in key countries in which primary forests are under pressure from growing populations, expansion of agriculture, poverty and commercial logging.

## FOREST HEALTH AND VITALITY

Most countries do not have reliable information on the area of forest affected by forest fires, insect pests, diseases and other disturbances such as weather-related damage because they do not systematically monitor these variables. For FRA 2005, only 20 countries reported on all four, and most of these were in Europe. At the global level, for countries that were able to report on different aspects, an average of 1.4 percent of their forest area was adversely affected by insects in an average year; 1.4 percent was affected by diseases; and 0.9 percent by forest fires. The data on other disturbances were not sufficient to draw conclusions at the global level.

There is a growing trend towards adopting more sustainable forest management strategies to contain forest pests, particularly in developed countries. These changes are related to changes in the perception and role of the forest, which is increasingly valued not just for economic reasons, but also for its ecological and social functions. In some regions, the risk of pests is being reduced – for example, through the replacement of large monocultures by smaller, mixed-species and mixed-aged stands in many European landscapes.

More information is available on pests of trees grown in developed than in developing countries – and also on pests in commercially valuable, planted forests rather than natural forests. Virtually nothing is known of the pests associated with trees harvested from natural forests in the tropics. Awareness of the need to gather and share information on forest pests is increasing, however; for example, 25 countries, including major forest countries such as Brazil, China and Indonesia, have provided information for a series of pest profiles (covering insects, diseases, nematodes, parasitic plants and mammals) currently being compiled by FAO.

Rapid transport, ease of travel and free trade have facilitated the spread of pests. In recent years, a number of invasive forest species have had an adverse impact on forestry and trade. For example, the movement of the Asian longhorn beetle, *Anoplophora glabripennis*, contributed to the adoption of an international standard for treating wood packaging material in international trade by the Interim Commission on Phytosanitary Measures of the IPPC.

A review of fire management based on papers prepared by national fire experts in different regions (FAO, 2006d) reached the following conclusions at the global level:

- An estimated 350 million hectares suffer wildland fires each year. This is equivalent to about 9 percent of total forest area, but the term "wildland" includes non-forest areas such as savannah, brush and open range. The actual damage to forests is less than 5 percent per year, but better data are needed.
- While many countries report that fire seasons are becoming more severe, there is insufficient information to conclude whether the total area burned or number of forest fires is increasing at the global level.
- At least 80 percent of fires are caused by people – and in some regions up to 99 percent. Agricultural needs and land clearing are the most common causes of fire,

**FIGURE 69** UN-ISDR/GFMC Global Wildland Fire Network

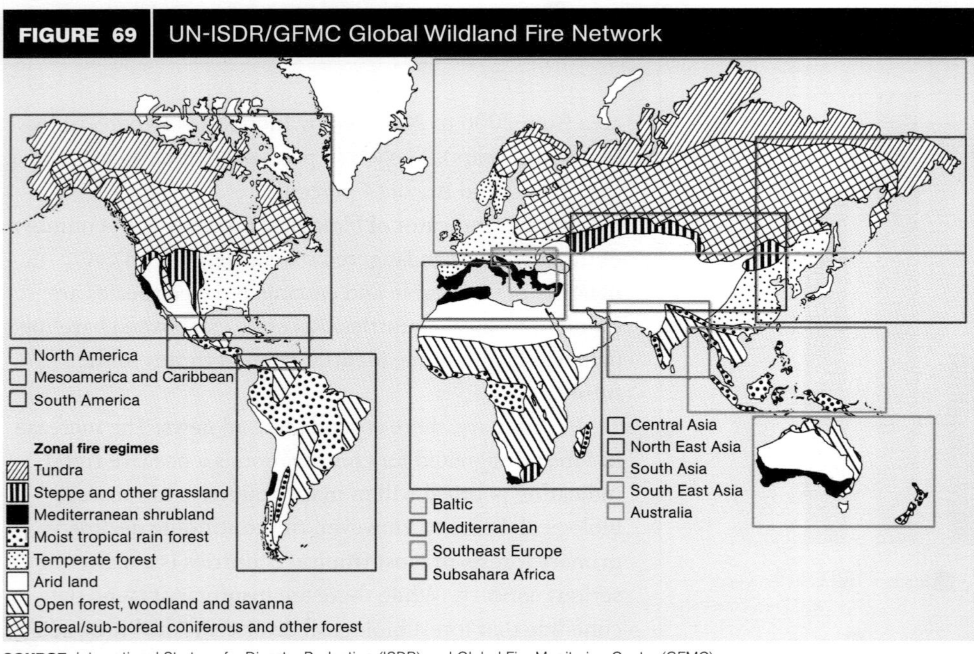


**SOURCE:** International Strategy for Disaster Reduction (ISDR) and Global Fire Monitoring Center (GFMC).

followed by arson. Lightning is the major non-human cause of wildfires.

- Fire can have positive or negative effects on forests, and its impact on forest health and vitality varies greatly in different ecosystems. Some countries experience almost no fires.
- Countries with serious wildfire problems have found that investing in fire prevention can be more cost-effective than concentrating on fire control, which is dangerous and expensive. Countries with fire management programmes invest in both approaches.
- Community-based fire management programmes are increasingly effective, in both developed and developing countries.
- International collaboration is increasing, as evidenced by the creation of 12 regional wildland fire networks (Figure 69) and approximately 100 transboundary bilateral fire agreements between neighbouring countries.

**FIGURE 71** Growing stock per hectare, 2005

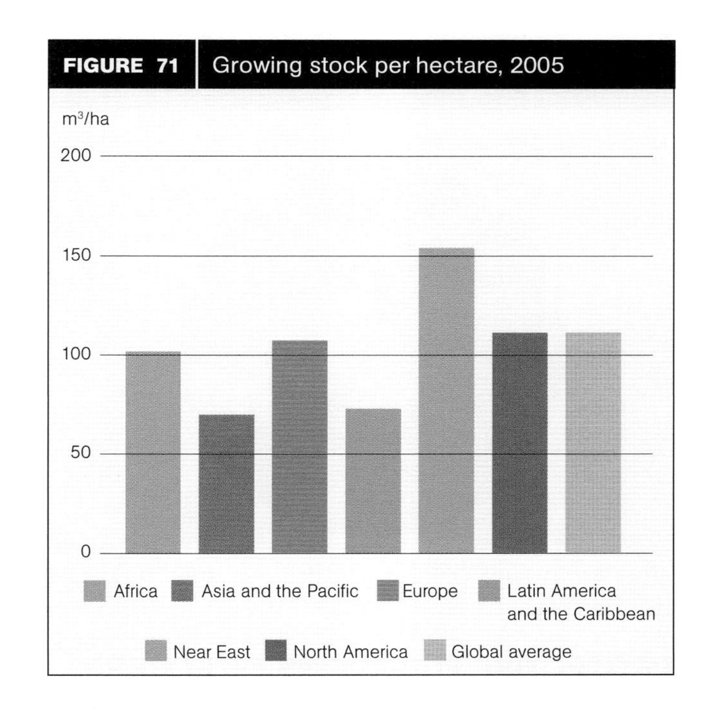

## PRODUCTIVE FUNCTIONS OF FOREST RESOURCES

As would be expected, the countries with the largest forest area represent the bulk of the world's wood volume (based on total growing stock) (Figure 70). However, growing stock per hectare varied among regions, owing primarily to climatic and other ecological differences (Figure 71).

In 2005, the area of forest designated for wood production as one of the management objectives was 50 percent. About 34 percent of the world's forest area was designated primarily for production. From 1990 to 2005, the forest area so designated declined by 5 percent, compared with a decrease in total forest area of 3 percent over the same period. This trend is not surprising, as the increase in forest area designated primarily for conservation of biological diversity increased by roughly the same amount. Perhaps this reflects a subtle change in global perceptions of forest values.

Countries that report a low percentage of their forest designated for production tend to report a high area for "multiple purpose", which usually includes production. It would appear that countries have different ideas on the meaning of this classification. For example, two of the largest producers of wood products in the world, the United States of America and Canada, report only 12 percent and 1 percent, respectively, as designated primarily for production.

The area of productive forest plantations increased by 2.5 million hectares between 2000 and 2005, indicating that a larger portion of wood removals may come from forest plantations in the future.

**FIGURE 70** Share in global wood volume in forests, 2005

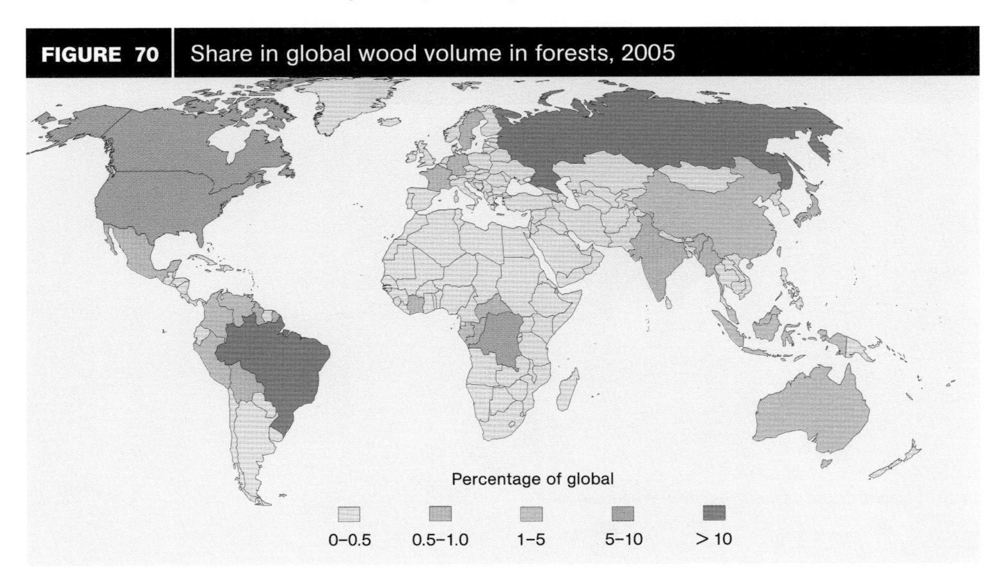

**FIGURE 72** Designated primary functions of forests, 2005

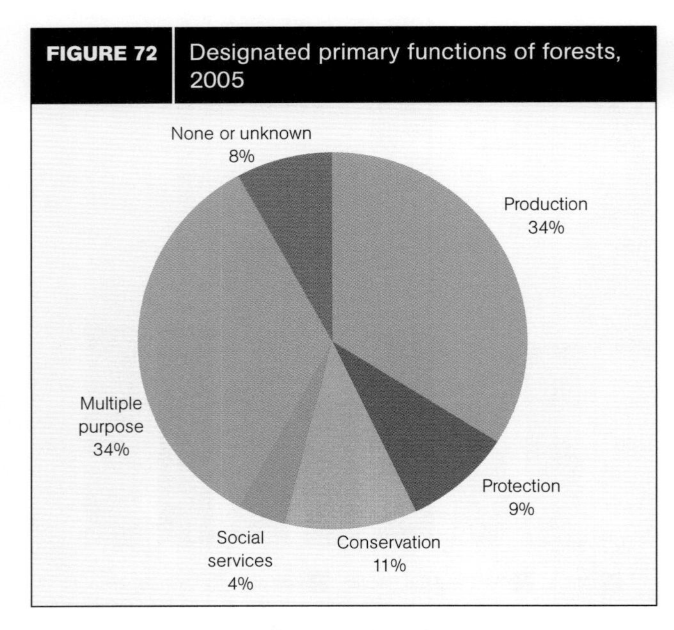


**FIGURE 73** Ten countries with largest area of protective forest plantations, 2005

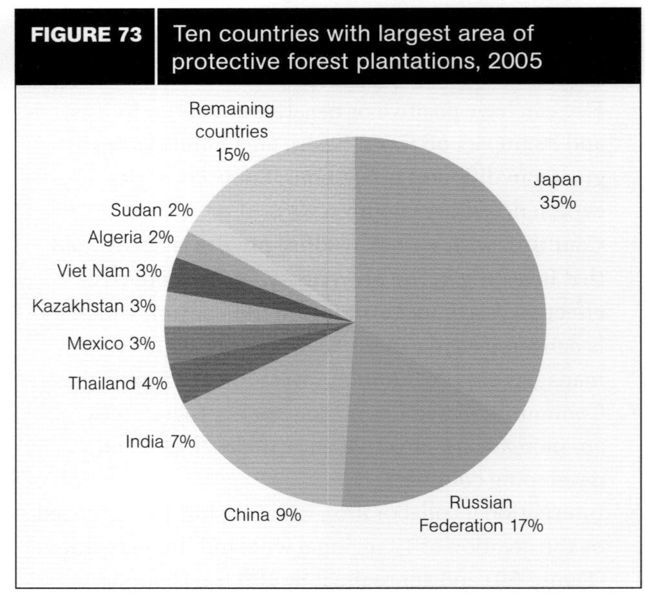


## PROTECTIVE FUNCTIONS OF FOREST RESOURCES

Some 9 percent of global forests are designated primarily for protection. However, not all countries use this category to classify their forests, and all forests perform some protective functions. Thus, while this is an interesting statistic, it clearly under-represents the extent of forests that perform protective functions. Additional research will be required to find improved variables to estimate this important forest function. In fact, a significant percentage of the world's forests are designated for multiple purposes, which can include forest protection (Figure 72).

In many countries, protective functions are the main reason for planting new forests or trees (Figure 73).

## SOCIO-ECONOMIC FUNCTIONS

The section on productive functions of forest resources presented information on global wood volume and growing stock (Figures 70 and 71). An illustration of the consumption of wood products completes the picture (Figure 74).

Primary forest products (roundwood, including industrial roundwood and fuelwood) represent a relatively large share of the forest sector's value in Africa, Asia and the Pacific and Latin America and the Caribbean. In contrast, wood-processing industries and pulp and paper account for the lion's share of the value of the sector in more developed regions.

Trade of forest products among countries is increasing (see Trade in forest products, p. 90). A positive net trade balance indicates that the value of exports exceeds the value of imports (Figure 75). Throughout the period from 1990 to 2004, Asia and the Pacific continued to be the major net importer of forest products. North America was, for many years, a net exporter, but in recent years it has become a net importer as well. The trend for Europe is the opposite of that for North America; today Europe is the leading net exporter of forest products.

**FIGURE 74** Consumption of wood products, 2003

**FIGURE 75** Trends in net trade of forest products by region

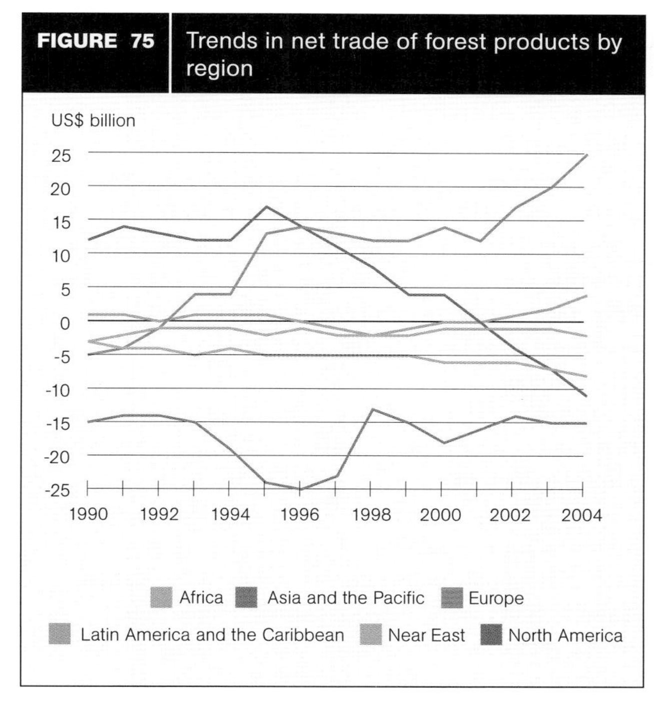


**NOTE:** A positive value indicates net export. A negative value indicates net import.

**FIGURE 76** Global trends in formal forest sector employment

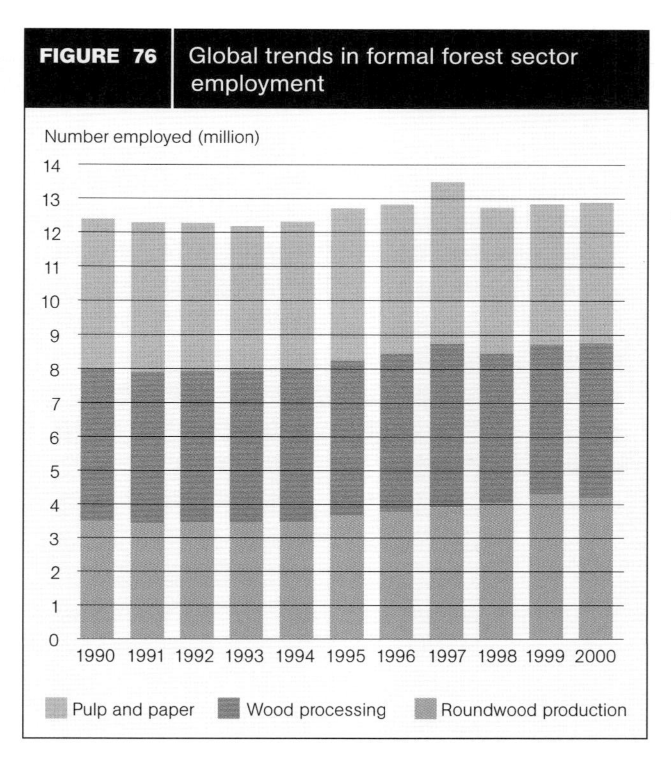


There was a gradually increasing trend in employment in the forest sector during the 1990s (Figure 76). At the global level, it is interesting to note that employment is roughly equal in the three major subsectors: roundwood production, wood-processing industries and pulp-and-paper industries. In general, roundwood production provides a larger share of jobs in developing countries, and the other two sectors provide most of the jobs in developed countries.

At the global level, employment increased by 4 percent from 1990 to 2000, while the forest sector's share of value added increased by only 1 percent (Table 37). These are significantly lower levels than in the global economy as a whole. Trade plays an increasingly important role in the forest sector, with exports continuing to increase much faster than other variables. Trade is especially significant in promoting economic growth in developing regions.

Almost 4 percent of the world's forests are managed primarily for social services such as recreation, education and tourism. Europe seems to give the most attention to social services provided by forests; almost 72 percent of Europe's forest area has social services as one of the designated functions.

TABLE 37
**Status and trends of forest-sector employment, value added and exports by region**

| Region | 2000 | | | | | | Change in absolute values, 1990–2000 | | |
|---|---|---|---|---|---|---|---|---|---|
| | Employment | | Value added | | Exports | | Employment | Value added | Exports |
| | (million) | (%) | (US$ billion) | (%) | (US$ billion) | (%) | (%) | (%) | (%) |
| Africa | 0.5 | 4 | 8 | 2 | 3 | 2 | 6 | 5 | 60 |
| Asia and the Pacific | 5.6 | 43 | 88 | 25 | 20 | 14 | 10 | −2 | 51 |
| Europe | 3.6 | 28 | 90 | 25 | 71 | 49 | −12 | −14 | 58 |
| Latin America and the Caribbean | 1.2 | 10 | 30 | 9 | 6 | 4 | 39 | 46 | 90 |
| Near East | 0.4 | 3 | 3 | 1 | <1 | <1 | 28 | −14 | 169 |
| North America | 1.5 | 12 | 136 | 38 | 44 | 31 | −1 | 10 | 33 |
| **All tropical countries** | **3.0** | **24** | **48** | **14** | **16** | **11** | **23** | **34** | **47** |
| **All temperate countries** | **9.9** | **76** | **306** | **86** | **128** | **89** | **−1** | **−2** | **50** |
| **World** | **12.9** | **100** | **354** | **100** | **144** | **100** | **4** | **1** | **50** |

**NOTE:** The changes in value added and exports are changes in the real value of these items (i.e. adjusted for inflation).
**SOURCE:** FAO, 2004a.

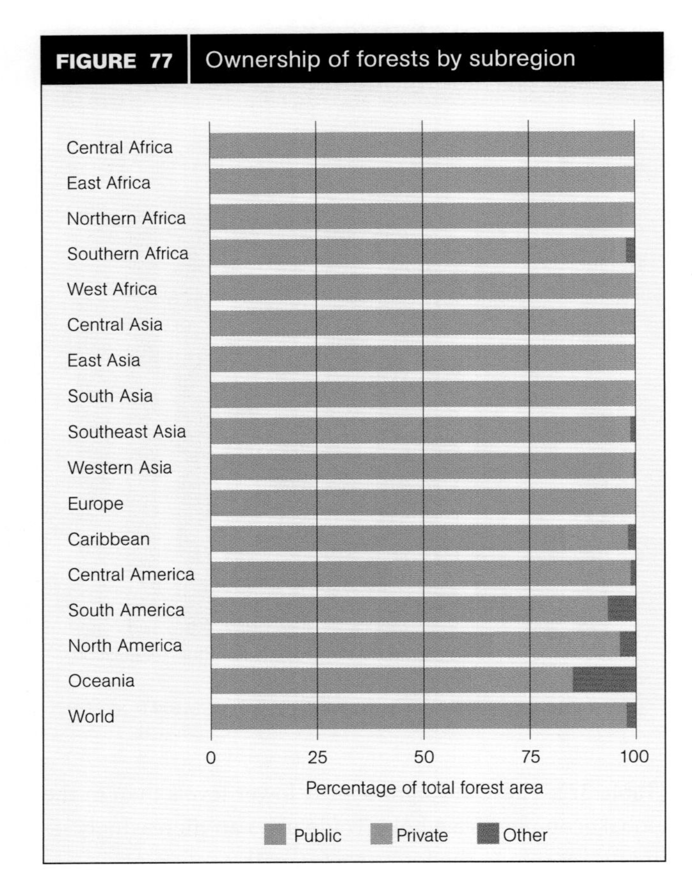


## LEGAL, POLICY AND INSTITUTIONAL FRAMEWORK

Legal, policy and institutional framework is perhaps the most important factor in setting the stage for sustainable forest management. Positive change is evident in all regions.

There are signs of political commitment towards sustainable forest management in the vast majority of countries. In the 15 years since the United Nations Conference on Environment and Development (UNCED), most countries have enacted new, more progressive forest laws and policies. Over 100 countries have established national forest programmes in an attempt to manage forests more holistically.

It is virtually impossible to compare the progress being made under this thematic element by country or by region. The nature of laws, policies and institutions is such that each country is unique. The first of the "Forest Principles" agreed to by all countries at UNCED was that management of forests is the sovereign responsibility of each country.

Nonetheless, several long-term trends are apparent. In many countries, forest lands are passing from national control to local management (devolution), although most forests remain under public ownership (Figure 77). In others, for example in Eastern Europe, there is a trend from public to private ownership (privatization). Awareness of the importance of secure forest tenure arrangements is increasing. In a number of countries, institutional responsibility for forests has shifted from agriculture ministries to environment ministries, reflecting a shift in emphasis from development towards conservation.

Despite the generally positive trends, much remains to be done. FAO, ITTO, the World Bank and bilateral aid agencies have a continuous backlog of requests from countries for assistance in strengthening forest policies and institutions. For example, FAO is able to undertake an average of about ten new projects each year to strengthen national forest institutions (through its Technical Cooperation Programme), but the demand from countries is considerably higher than the ability to respond. The National Forest Programme Facility supports the efforts of more than 40 countries to increase the participation of all stakeholders in forest decision-making processes (Box 4), but the demand for additional assistance far exceeds its capabilities.

## BOX 4 National Forest Programme Facility

The National Forest Programme Facility (see www.fao.org/forestry/site/30766/en) is a funding mechanism hosted by FAO that supports active stakeholder participation at the country level in the development and implementation of national forest programmes. It focuses on capacity-building and information-sharing, and offers information services on national forest programmes worldwide.

The Facility stimulates participation in the national forest programme process by providing grants directly to stakeholders in partner countries through a competitive and transparent process. Its overall objective is to assist countries in developing and implementing national forest programmes that effectively address local needs and national priorities and reflect internationally agreed principles. Informed and broad participation is the key to achieving this objective.

Since it was created in 2002, the Facility has developed partnerships with 42 countries and four subregional organizations; it has allocated US$6 million under 220 grants to stakeholders, about 70 percent of which are non-governmental (see figures). Facility grants have supported the participation of stakeholders in formulating policies and strategies, broadening national forest programmes and developing new legal, fiscal and institutional instruments. The Facility has also launched information-sharing initiatives.

Country support, 2002–2006

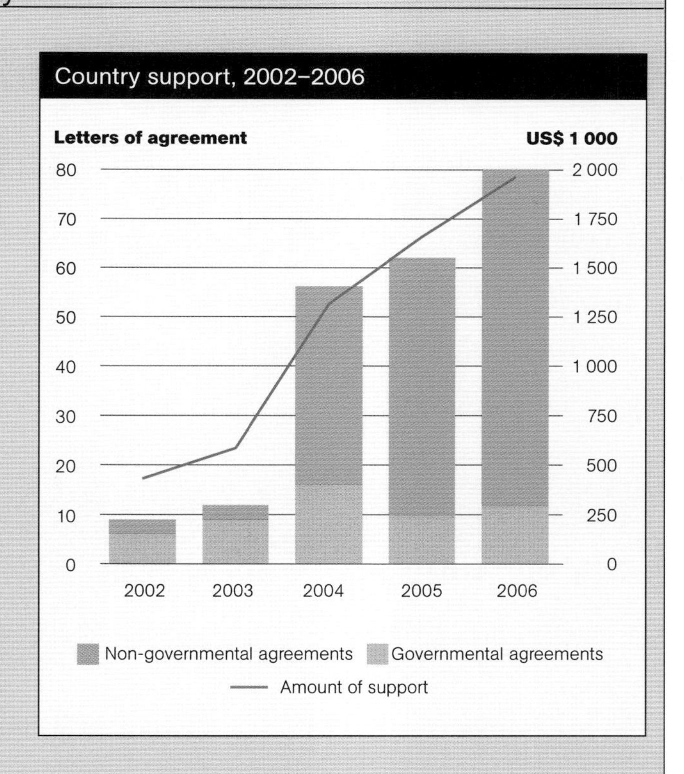


National Forest Programme Facility partners

PART 2

# Selected issues in the forest sector

FAO/FO/C. Jekkel

**THERE IS INCREASING EVIDENCE** that forests will be profoundly affected by climate change. The recent outbreak of the mountain pine beetle in British Columbia, for example, appears to be related to historically high temperatures and may become the worst forest catastrophe in Canadian history.

On the other hand, forests can play a key role in mitigating climate change. However, the world is struggling with political and bureaucratic hurdles that are limiting the use of the Kyoto Protocol (United Nations, 1998) as an instrument to help stop tropical deforestation.

After its entry into force in February 2005, implementation of the Protocol and its mechanisms is slowly gaining momentum, but there has been little impact in the forest sector. As of 2006, 25 methodologies for setting baselines and monitoring Clean Development Mechanism (CDM) projects (in all categories) had won approval, and 64 projects employing one of the approved methodologies had been registered. Many more projects are in the pipeline (Figure 78).

Forestry projects lag behind those of other sectors (Figure 79). Among the hurdles is the decision by the European Commission not to admit carbon credits from forestry projects in its internal emission trading scheme. However, individual European Union governments are free to purchase such credits, and a review of the Commission's decision was expected in late 2006.

**FIGURE 78** Increase in the number of CDM projects under development in 2005

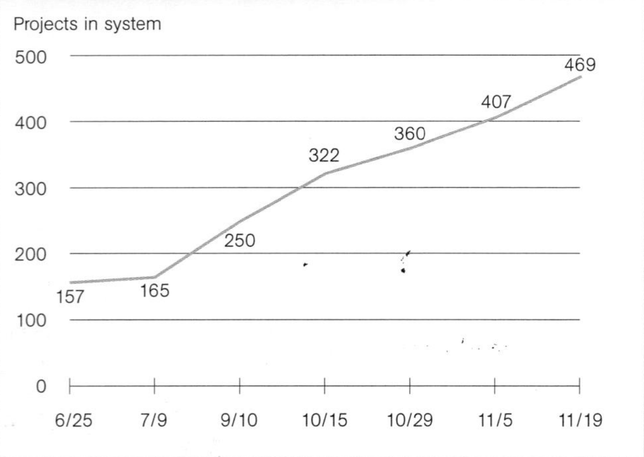

CDM rules have been modified to allow the bundling of small- or large-scale projects. This opens up the possibility of relatively risk-free, small-scale afforestation and reforestation projects, and will facilitate the involvement of low-income individuals and communities.

Climate change negotiations have tended to focus on greenhouse gas emissions in industrialized countries. But attention now also encompasses developing countries, whose emissions are substantial and increasing. Attention focuses in particular on the role played by deforestation – which causes 35 percent of emissions in developing countries and fully 65 percent in the least developed countries. Unusually high participation in the United Nations Framework Convention on Climate Change (UNFCCC) Workshop on Reducing Emissions from Deforestation in Developing Countries, held in Rome, Italy, from 30 August to 1 September 2006, was a clear sign that developing countries are ready to begin reducing their emissions from land-use changes and that the climate-change regime is furthering its role in the global effort to reduce deforestation. Financing is a key hurdle. The workshop proposed several new mechanisms for transfer of payments from developed to developing countries. Negotiations will continue at a second workshop to be held in 2007.

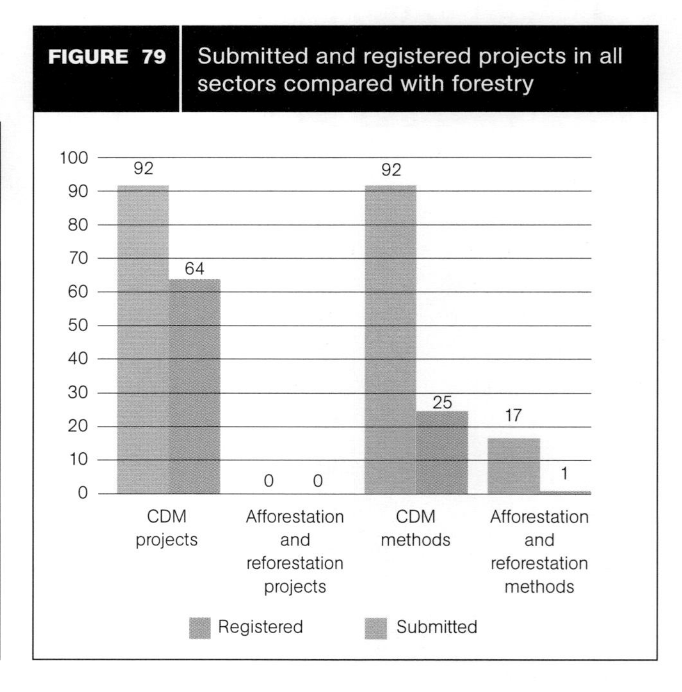

**FIGURE 79** Submitted and registered projects in all sectors compared with forestry

**THE YEAR 2006** was the International Year of Deserts and Desertification (IYDD), declared by the United Nations General Assembly to raise public awareness of this crucial issue. Observation of IYDD was led by an interagency committee of partners active in the implementation of the United Nations Convention to Combat Desertification (UNCCD), including the United Nations Environment Programme (UNEP), the United Nations Development Programme (UNDP) and FAO. Countries and civil society groups organized international events and special initiatives such as tree-planting ceremonies to spread the message that desertification is a global problem, thus helping to strengthen the place of dryland issues on the international environment agenda.

Desertification constitutes one of the world's most alarming processes of environmental degradation. It affects about two-thirds of the countries of the world, more than one-third of the earth's surface (more than 4 billion hectares) and more than one billion people, with potentially devastating consequences on livelihoods and food security.

Desertification refers to the degradation of land resulting from various forces, including climatic variability and unsustainable human activities such as overcultivation, overgrazing, deforestation and wildland fire. It reduces the biological and economic productivity of land and has negative effects on rivers, lakes, aquifers and infrastructure. Desertification reduces food security and can lead to social unrest and conflict. By the year 2020, an estimated 135 million people risk being driven from their lands because of continuing desertification, including 60 million in sub-Saharan Africa alone.

With the world's highest rate of desertification, sub-Saharan Africa is facing losses in the productivity of cropping land approaching 1 percent annually and has had a productivity loss of at least 20 percent over the last 40 years (World Meteorological Organization, personal communication, 2006). Other affected areas of the world include one-quarter of Latin America and the Caribbean and one-fifth of Spain. In China, sand drifts and expanding deserts have taken a toll since the 1950s of nearly 700 000 ha of cultivated land, 2.35 million hectares of rangeland and 6.4 million hectares of forests, woodlands and shrublands. Worldwide, some 70 percent of the 5.2 billion hectares of dry lands used for agriculture are already degraded and threatened by desertification.

In spite of the social and environmental impacts of desertification, there is no updated information on the progression of this process. The World Bank has been using the same estimate of annual losses resulting from desertification (US$42 billion) since 1990 (World Bank, 2006a). The Millennium Assessment is based on a model developed in the early 1980s. Updated information is a key factor for more effective action. UNEP estimates that an effective 20-year global effort to combat desertification would cost in the range of US$10 to 22 billion per year.

Natural vegetation plays a fundamental role in fighting soil degradation, and perennial vegetation guarantees effective and long-lasting soil protection. Deforestation increases the vulnerability of land to desertification. Afforestation and reforestation, within an appropriate landscape approach, are among the most effective ways to counteract it.

The financing of efforts to halt desertification is perhaps the most problematic issue facing countries with low forest cover. Many of the countries are poor and are already facing difficulties in repaying loans to international financial institutions. The World Bank, regional development banks and UN organizations and agencies have a role, as well as the Global Mechanism for UNCCD. Since land degradation was adopted as one of its focal areas, the Global Environment Facility is a potential source of funds, as well.

**Green belt outside Nouakchott, Mauritania, protects the city from moving sand dunes: left, mechanical sand dune fixation precedes tree planting; right, established protection stands (2005)**

**FOREST LANDSCAPE RESTORATION** brings people together to identify, negotiate and implement practices that restore an optimal balance among the ecological, social, cultural and economic benefits of forests and trees, within the broader pattern of land uses. It involves practical approaches that do not try to re-establish the pristine forests of the past. Rather, the goal is to adopt holistic approaches that restore the functions of forests and trees and enhance their contribution to sustainable livelihoods and land uses.

The Global Partnership on Forest Landscape Restoration (see www.unep-wcmc.org/forest/restoration/globalpartnership) is a worldwide network of more than 25 governments and organizations working to strengthen forest landscape restoration efforts globally (Box 5). Partners share their expertise with other practitioners, governments, communities and businesses. Several members of the Collaborative Partnership on Forests (CPF) are also active in this partnership.

Phase I (2003–2005) of the global partnership focused on raising the profile and understanding of forest landscape restoration, establishing national working groups, securing funds and providing technical support to its advocates. An International Forest Landscape Restoration Implementation Workshop was held 4–8 April 2005 in Petropolis, Brazil.

Forest landscape restoration – a mosaic of planted forests for wood production and secondary naturally regenerated forests for protection of valleys and waterways, Bahia State, Brazil

Phase II (2005–2009) aims to: increase the partnership; extend the network of learning sites to improve understanding and practices in forest landscape restoration; encourage broader, multistakeholder participation; strengthen the legal, policy, regulatory and institutional frameworks for forest landscape restoration; provide critical information and tools for sound development; and host a second international workshop.

Forest landscape restoration can be a vehicle to deliver on commitments regarding forests, biodiversity, climate change and desertification, as well as contributing towards achieving the Millennium Development Goals. It involves a multidisciplinary approach that integrates policies, plans and practices across sectors in national development processes, including eradicating hunger, reducing poverty and managing natural resources on a sustainable basis. Consequently, it also involves the integration of national forest programmes, policies and plans into national development programmes.

Experience has shown that successful forest landscape restoration starts from the ground up, with people who live in the landscape and stakeholders affected by its management. There is no single blueprint for success, as each situation depends upon unique local circumstances and processes. Examples of forest landscape restoration in the pattern of land uses interspersed in the landscape may include:

- management of natural forests for protective functions (e.g. watershed management or conservation of biodiversity);
- management of natural forests for productive functions (e.g. wood, fibre and NWFPs);
- soundly planned and managed planted forests created through afforestation and reforestation;
- riparian buffer zones along watercourses for flood and erosion protection;
- retention of forested corridors between remnant forest areas;
- rehabilitation of secondary forests;
- management of natural forests for tourism and recreation;
- agroforestry, combining trees, agricultural crops and livestock;
- community-based forest and tree development in phase with people's needs;
- urban and peri-urban forestry.

**BOX 5** Members of the Global Partnership on Forest Landscape Restoration

World Conservation Union (IUCN)

World Wide Fund for Nature (WWF)

Alliance of Religions and Conservation (ARC)

CARE International

Secretariat of the Convention on Biological Diversity (CBD)

Centre for International Forestry Research (CIFOR)

Food and Agriculture Organization of the United Nations (FAO)

Global Mechanism for the UN Convention to Combat Desertification

World Agroforestry Centre (ICRAF)

International Union of Forest Research Organizations (IUFRO)

International Tropical Timber Organization (ITTO)

Program on Forests (PROFOR)

Secretariat of the United Nations Forum on Forests (UNFF)

UNEP World Conservation Monitoring Centre (UNEP-WCMC)

Forestry Research Institute of Ghana (FORIG)

Governments of El Salvador, Finland, Italy, Japan, Kenya, Lebanon, South Africa, Switzerland, the United Kingdom of Great Britain and Northern Ireland and the United States of America

# Forestry and poverty reduction

**THE CROSS-SECTORAL** and participatory nature of national forest programmes makes them ideal mechanisms for gathering and sharing information from a wide range of sources on country issues, priorities and initiatives both within and outside forestry. As such, they can be instrumental in addressing the marginalization and underfunding of the sector by building linkages to wider national agendas, including poverty reduction and similar development strategies. However, the literature suggests that ties are often weak or non-existent.

Interviews in 2005 with government authorities and NGOs in Namibia, the Niger, Nigeria, the Sudan, Tunisia, Uganda, the United Republic of Tanzania and Zambia explored the extent to which national forest programmes and other sectoral processes were linked to poverty reduction strategies or similar frameworks. Best practices and constraints on and opportunities for establishing effective linkages were identified, drawing on lessons from forestry and other sectors such as agriculture, energy, health and education. The following findings are indicative of trends.

- Poverty reduction as a national goal is sharpening the focus on cross-cutting issues.
- Governments are instituting sector-wide approaches to planning and resource allocation.
- Efforts to assess and report on poverty are increasingly involving stakeholders, but participatory processes are often time-consuming and costly.
- Decision-makers will continue to underestimate the importance of forestry to social and economic development as long as the sector fails to quantify the full extent of its contributions, including those from fuelwood, NWFPs and environmental services.
- Change in donor funding from sectoral to central support may weaken forestry capacity and hinder efforts to decentralize services to districts and communities where interventions could have the greatest potential to alleviate poverty.
- Marketing forestry on the basis of its capacity to meet key objectives of other sectors will broaden understanding of the benefits and open up opportunities for collaboration.

In partnership with IUCN, the Overseas Development Institute, the Center for International Forestry Research (CIFOR) and Winrock International, the Program on Forests (PROFOR) is attempting to show how sustainable forest management can enhance rural livelihoods, conserve biological diversity and help achieve the Millennium Development Goals. Under this partnership, case studies were carried out in Guinea, Honduras, India, Indonesia, the Lao People's Democratic Republic, Mexico, Nepal and the United Republic of Tanzania. In addition, a poverty–forests linkages toolkit was developed to increase understanding of how forests contribute to livelihoods (see www.profor.info/content/livelihood_poverty.html). The material includes:

- methods for gathering information on economic and other contributions from forests to households, especially those of poor people;
- field-data analyses that determine how forests can reduce poverty and vulnerability;
- suggestions for a packaging of results that is relevant to local and national planners, governments, institutions and organizations;
- description of poverty reduction strategy processes, including potential entry points for forestry, and an indication of the skills required to influence outcomes;
- case studies that illustrate the contributions of forest resources to households and an analysis of the impact of forestry policies and programmes.

**Development organizations are increasingly working to show how sustainable forest management can contribute to poverty reduction and to strengthen links between forestry programmes and poverty-reduction strategies**

# Forestry sector outlook

**RECENT DECADES** have witnessed rapid socio-economic changes affecting all aspects of life, including the society/forest relationship. Globalization, accelerated by the rapid expansion of information and communications technologies, has brought countries and people together, and the ease of movement of capital and technology has altered the economic landscape. At the same time, the unevenness of globalization has excluded a large number of people from realizing the potential benefits. Society is grappling with change-related environmental problems, including loss of biological diversity, land degradation and desertification, climate change and increasing costs of energy and water.

An understanding of how the society/forest relationship is likely to evolve is important in preparing the sector to address emerging challenges and opportunities. Strengthening country-level strategic planning requires a better understanding of the developments beyond national borders. It is in this context that FAO is implementing regional and global forestry sector outlook studies.

Regional forestry sector outlook studies are carried out on a rotating basis. Of the five studies, the one on Asia and the Pacific was completed first (in the late 1990s), followed by Africa, Europe, Latin America and the Caribbean and Western and Central Asia, and a new study is currently under way to extend the Asia and the Pacific outlook from 2010 to 2020.

Global studies build on regional studies and are available on specific themes, including: a global fibre supply model; global forest products consumption, production, trade and prices; and the global outlook for wood supply from forest plantations. Most of the current projections extend through 2010. By the end of 2007, these projections will have been extended through 2030.

Global and regional outlook studies synthesize information from a variety of sources to provide a coherent view of the overall direction of change (Box 6). The focus is on analysing driving forces and how they directly and indirectly have an impact on forests. "What happens to forests" will be largely determined by "what happens outside forests".

**BOX 6** What does the future hold for forests and forestry?

Drawing upon FAO global and regional outlook studies, a number of trends have been observed:

- Deforestation and forest degradation will continue in most developing regions; a reversal of the situation would depend on structural shifts in economies to reduce direct and indirect dependence on land. In most developing tropical countries, agricultural land used for both subsistence and commercial cultivation continues to expand. Consequently, loss of forests will continue.
- In contrast, deforestation has stopped in countries where the agricultural land base has shrunk. Continued expansion of forests is expected in parts of Asia and the Pacific, Europe and North America. A shift away from fossil fuels and towards biofuels will have divergent impacts, in some cases resulting in expansion of forests while, in others, continued degradation. However, the reduced economic viability of forestry may result in lower investments in forest management.
- The possible effects of climate change may increase the incidence and severity of forest fires and pest and disease infestation and may alter forest ecosystems. At the same time, concern about climate change will focus increased attention on the role of forests in carbon conservation and sequestration and in substitution of fossil fuels.
- Increasingly, forests will be valued for their environmental services. The protection of biodiversity and the arresting of desertification and land degradation will assume greater importance.
- Recreational use of forests is receiving more attention, especially in developed and rapidly developing countries, requiring changes in the approach to forest management.
- Technological changes, such as biotechnology and materials technology (especially engineered wood), will improve productivity and reduce raw material requirements.
- Geographical shifts in production and consumption are likely to intensify, especially as a result of rapid growth of emerging economies in Asia and the Pacific and Latin America and the Caribbean. This will be countered by slow growth of demand in many developed countries, owing to demographic changes and lower income growth rates.
- For many developing countries, wood will remain the most important source of energy. The rising price of oil and increasing concern for climate change will result in increased use of wood as fuel in both developed and developing countries. The development of improved fuel conversion technologies that enhance energy efficiency would particularly favour this shift.

# Forest tenure

**THERE HAVE BEEN** many projects, workshops, case studies and reports on participatory and community forestry over the past 30 years, but is there quantitative evidence of real change? One such measure would be the extent to which forest ownership and management rights have devolved to local communities or to individuals. This could take many forms:

- recognition of ownership or tenure of forest land by community groups;
- devolution of the management of selected state forest areas to local users' groups;
- joint management or co-management of state forest lands;
- leasing of state lands for forestry purposes;
- community concessions.

Public forest ownership remains by far the predominant category in all regions (FAO, 2006a). At the global level, 84 percent of forest lands and 90 percent of other wooded lands are publicly owned. The area of forests owned and administered by communities doubled from 1985 to 2000 – reaching 22 percent in developing countries – and is expected to increase further (White and Martin, 2002).

A 2005 study of forest tenure in 19 countries in Southeast Asia (FAO, 2006j) revealed that about 365 million hectares of forests (92 percent) are public, the majority of which (79 percent) are owned by central governments (Figure 80). The percentage owned by local communities and groups and indigenous peoples is insignificant. Most public forests (63 percent) are managed directly and solely by central or local governments. However, when considering forests owned or managed by local forest holders, this area increases to 18 percent of total forest area. Short-term agreements, with a limited devolution of management rights and responsibilities, prevail over longer, more secure tenure agreements.

**FIGURE 80** Forest ownership in selected countries in Southeast Asia

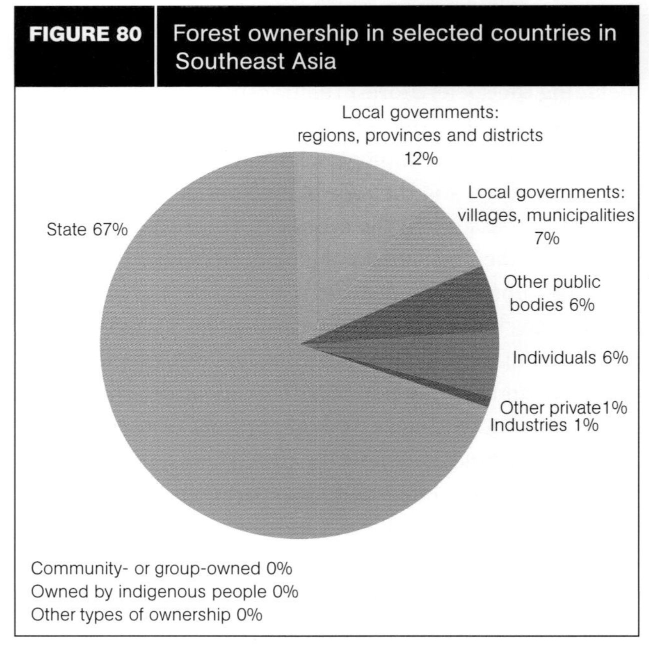


**SOURCE:** FAO, 2006j.

The transfer of forest management and user rights often is not accompanied by adequate security of tenure and the capacity to manage these resources. Improving local rights and access to forest resources are prerequisites to sustainable forest management. Much remains to be done to secure these rights and to remove policies and institutional frameworks that hamper the scaling-up of participatory forestry. Understanding the impact of forest tenure is essential if governments are to formulate effective policies and promote sustainable use and stakeholder participation.

**LIKE ANY OTHER** industrial activity, harvesting of wood and non-wood products has an impact on the natural and social environment. Reduced-impact logging methods such as low-intensity, selective harvesting cause minimal environmental damage and are economical if environmental impacts such as damage to residual stands are factored in (FAO, 2004b).

Yet inappropriate harvesting methods are still widely used throughout the tropics – to the detriment of the well-being of the workforce and local population, environmental sustainability and efficiency. Damaging practices may include:

- hyper-selective harvesting or skimming, which jeopardizes polycyclic harvesting systems and tempts foresters to re-enter harvested blocks;
- failure to implement a harvesting plan;
- inadequate road planning and construction;
- uncontrolled felling and wasteful topping and trimming;
- causing excessive crawler tractor skid trails on vegetation and soil, rather than working only on marked trails when carrying out skidding operations;
- wasteful conversion of timber owing to inappropriate topping and grading at the landing;
- lack of monitoring, control and impact assessment.

Environmentally friendly harvesting: skidding of Okume ties on a preplanned skid trail, without damage to soil or residual vegetation, Gabon

Why do these practices persist? The reason is a combination of unawareness and economics. Many companies or individuals in the harvesting business are unfamiliar with reduced-impact logging practices; they do not realize that such practices are as economically viable in the long run as traditional ones; contractors are not trained; more destructive logging and road-building practices are in place; some simply do not care. Unfortunately, many loggers think in short time frames and do not consider the environmental impacts. In many instances, companies have no legal or apparent financial incentive to improve their harvesting practices.

While intergovernmental processes have struggled to make progress at the policy level, considerable progress has been achieved at the field level regarding steps to be taken to manage forests sustainably. The *FAO model code of forest harvesting practice* was developed in 1996 (FAO, 1996). Regional codes were subsequently agreed in Asia and the Pacific in 1999 and in West and Central Africa in 2003 and 2005. National-level codes have been adopted or are under preparation in several countries in Southeast Asia. Parallel to the publication of the codes, implementation strategies were developed and training and activities increased in many countries, but progress in the field remains slow.

Two main hurdles threaten successful implementation of best practices in forest harvesting: illegal logging practices, which are undermining fair market conditions, and widespread lack of awareness of or concern for the economic, environmental and social benefits of good harvesting practices.

Forests and people suffer from destructive and wasteful harvesting practices. Technical, social and environmental guidelines are available, but are insufficiently implemented. More awareness-raising, training and research are needed to overcome these challenges.

# Invasive species

**"INVASIVE SPECIES"**, also known as "alien species" or "alien invasive species", are species whose introduction, establishment and spread into new areas threaten ecosystems, habitats or other species and cause social, economic or environmental harm, or harm to human health. Invasive species can be found in all taxonomic groups, from bacteria to mammals, and are second only to habitat destruction as a threat to global biodiversity (Mooney and Hofgaard, 1999).

Many factors can support the introduction and spread of invasive species, including land-use changes, forest activities (harvesting of wood and NWFPs, forest road construction and conversion of natural forest to planted forests), tourism and trade.

Particularly challenging to forest managers are non-native tree species that have been intentionally introduced into an ecosystem to provide economic, environmental or social benefits (Figure 81). Many tree species used for agroforestry, commercial forestry and desertification control are not native to the area. It is vital to ensure that such species serve the purposes for which they were introduced and are managed so as not to cause negative effects on native ecosystems (see FAO, 2005b).

Estimates of the full costs of biological invasions are rare because of the difficulty in assessing the costs of impacts on biodiversity, ecosystem functions and human health, or other indirect costs such as the impact of control measures. The costs of invasive species to the forest sector have not been studied on a global scale. However, based on a study of six countries (Australia, Brazil, India, South Africa, the United Kingdom and the United States of America), it was estimated that as many as 480 000 alien species have been introduced in agriculture and forestry worldwide, with an annual cost of more than US$1.4 trillion (Pimentel *et al.*, 2001).

Preventing and reducing the harmful effects of invasive species requires an approach that incorporates biological, ecological and social sciences, economics, policy analysis and engineering. National efforts should include early warning systems, eradication and control, as well as increased awareness and political leadership. Global, regional and bilateral efforts include standards and guidelines, monitoring and assessment, and information and action networks.

Numerous international and regional programmes and instruments, binding and non-binding, have been developed to address the problem of invasive species, some with direct or indirect implications for forests and the forest sector.

The Convention on Biological Diversity (CBD), for example, calls on its parties to "prevent the introduction of, control or eradicate those alien species which threaten ecosystems, habitats, or species" (Article 8[h]). The parties have adopted a series of 15 guiding principles to lead governments and organizations in developing effective strategies for minimizing the spread and impact of invasive alien species. The eighth Conference of the Parties (COP-8) to CBD, held 20–31 March 2006 in Brazil, focused on gaps and inconsistencies in the international regulatory framework addressing invasive species. A process for the in-depth review of invasive species is scheduled for COP-9 in 2008.

The Global Invasive Species Programme (GISP) was established in 1997 to address global threats caused by invasive species and to provide support to the implementation of Article 8(h) of CBD. To increase awareness and provide policy advice, the Programme has prepared the *Global strategy on invasive alien species*, which outlines ten strategic responses to the invasive species problem (McNeely *et al.*, 2001). GISP initiated the Global Invasive Species Information Network (www.gisinetwork.org), a Web-based network of governmental, non-governmental, educational and other organizations working together to provide increased access to data and information on invasive species worldwide.

In addition, FAO and partner countries have recently created two regional networks: the Asia–Pacific Forest Invasive Species Network (APFISN) (www.fao.org/forestry/site/35067/en) and the Forest Invasive Species Network for Africa (FISNA) (www.fao.org/forestry/site/26062/en).

Because invasive species are addressed in different agreements and conventions (e.g. CITES, IPPC and the World Trade Organization Agreement on the Application of Sanitary and Phytosanitary Measures), many countries have difficulty in keeping up with the reporting requirements. In response, the UNEP World Conservation Monitoring Centre (UNEP-WCMC) has recently developed a set of issue-based modules summarizing country obligations under CBD and other conventions, which facilitates more streamlined and efficient delivery on reporting obligations (see svs-unepibmdb.net/?q=node/14).

FIGURE 81 Introduction and spread of non-native woody species

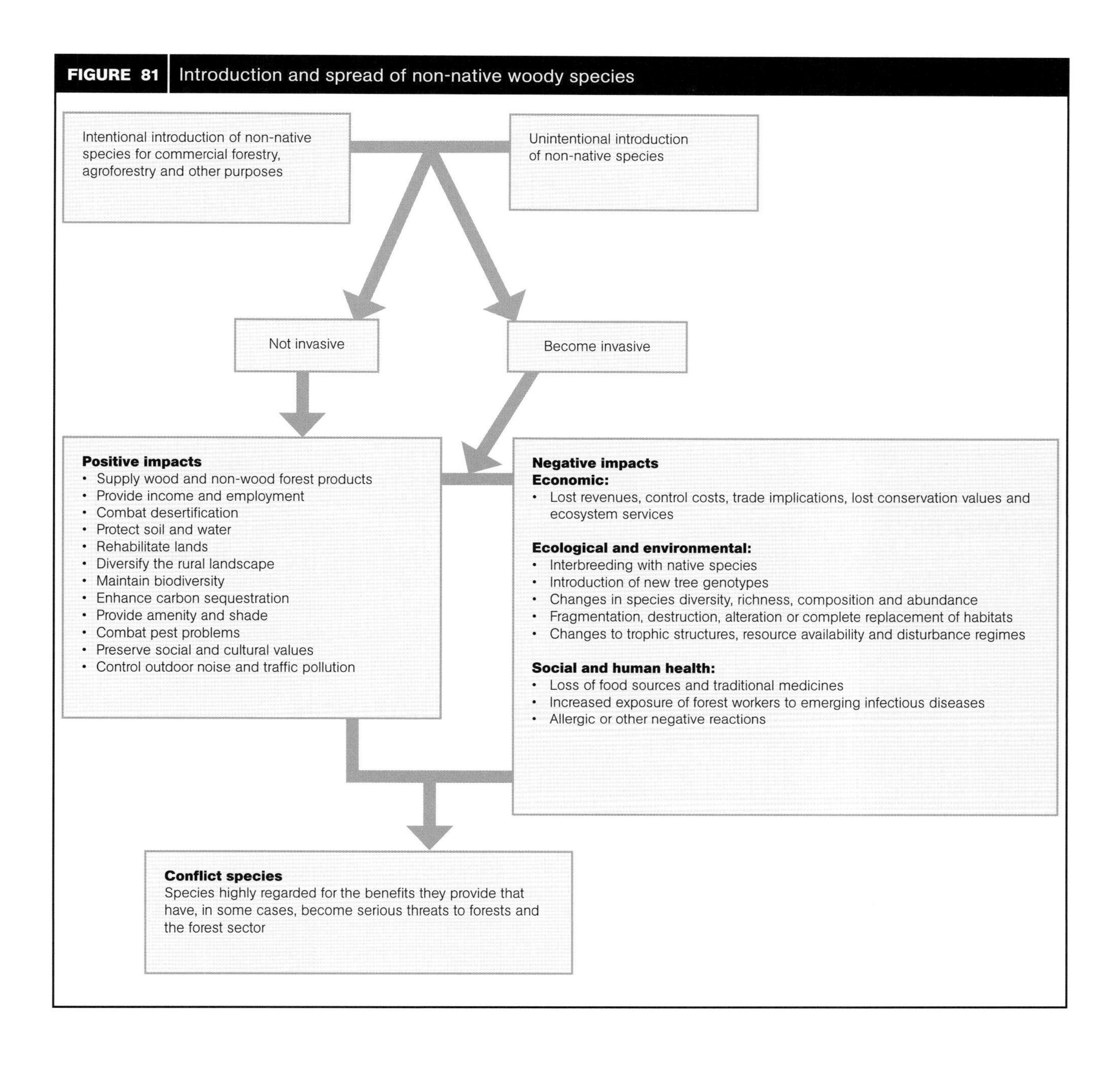

# Monitoring, assessment and reporting

**THE IMPORTANCE OF** monitoring, assessment and reporting on forests has gained widespread attention in the international forestry community. In recent years, progress has been made in several key areas.

## MEASURING PROGRESS TOWARDS SUSTAINABLE FOREST MANAGEMENT

Many regional processes, FAO and ITTO have made contributions to monitoring, assessment and reporting for sustainable forest management. Fifteen years of work on criteria and indicators regionally and nationally have contributed to a common understanding of this concept. One result is that the Global Forest Resources Assessments continue to improve in terms of the scope of coverage, data quality and country participation (Box 7).

## CAPACITY-BUILDING TO IMPROVE THE NATIONAL INFORMATION BASE

Steady progress is being made in strengthening the capability of countries to carry out monitoring, assessment and reporting, but it is hindered by a shortage of resources. The estimated cost of carrying out a one-time national forest assessment, based on relatively low-intensity, systematic field sampling, is from US$500 000 to US$1 million, depending on the country. In the past five years, FAO has supported the preparation of national forest assessments in 14 countries; in addition, three regional projects are planned or under way. This is a good start, but another 100 countries require assistance. The projects aim to strengthen national capabilities to monitor, assess and report on forest resources, products and institutions. ITTO supports training in the use of its criteria and indicators and related reporting format.

### Status of national forest assessments

**Assessment completed:** Cameroon, Costa Rica, Guatemala, Lebanon, Philippines

**Assessment in progress:** Bangladesh, Congo, Honduras, Zambia

**Assessment formulated:** Cuba, Kenya, Kyrgyzstan, Nigeria

**Assessment under formulation:** Viet Nam, regional project in West Africa (9 countries), regional project in the Near East (7 countries), regional project in Southern Africa (SADC countries)

**Project in progress:** Regional project on monitoring, assessment and reporting in Asia (FAO/Japan)

## INTERNATIONAL COMMITMENT

The World Resources Institute is working in selected countries to develop a forest mapping tool, based on satellite imagery, to help enforce laws and monitor illegal operations. In its work to streamline forest-related reporting to international processes, CPF is developing a common information framework that will improve information management and reduce the reporting burden on countries. It has provided easy access to national reports submitted to CBD, FAO, ITTO, UNFCCC, UNFF and other forest-related reporting processes and is working towards more-coordinated, unified information requests (www.fao.org/forestry/cpf-mar).

Progress has been made in harmonizing forest-related definitions though a series of expert meetings. A common understanding has been reached on the definitions of: forest, forest degradation, rehabilitation, restoration, fragmentation, natural forest, planted forest, forest plantation, forest management and managed forests. Despite this progress, some forest terminology remains inconsistent, and new definitions arising during international processes make it difficult to monitor trends in variables.

## FUTURE CHALLENGES

- Forest data are weak in many countries. Information gaps make it difficult for countries to arrive at sound policy decisions and to implement sustainable forest management, including effective law enforcement.
- New technologies may improve the availability and reduce the cost of high-resolution satellite images for the monitoring of deforestation, forest degradation, shifting cultivation, biomass, growth and yield, and other useful variables. However, few countries have the resources to use these capabilities.
- Quality information requires a long-term investment.
- International organizations need to focus on acquiring information that is truly useful to member countries. Too many long questionnaires are circulated by too many organizations.
- CPF organizations need to expand their efforts to streamline reporting, to eliminate duplication and to present information in a consistent manner.
- Knowledge that is shared is powerful and cost-effective. Countries and organizations need to explore new partnership approaches.

## BOX 7 Kotka V Expert Consultation on Global Forest Resources Assessment: Towards FRA 2010

The Global Forest Resources Assessment programme has received technical guidance from international specialists through expert consultations organized by FAO and UNECE at regular intervals over the last 20 years. The first consultation was held in 1987, and subsequent ones took place in 1993, 1996 and 2002. The most recent consultation, the fifth, was held 12–16 June 2006.

All consultations have been hosted by the Finnish Forest Research Institute (Metla) and have been held in the city of Kotka, Finland. Thus the most recent consultation is referred to as Kotka V.

Kotka V had two main objectives:

- to provide guidance for FRA 2010 based on an in-depth evaluation of FRA 2005; and
- to enhance collaboration with other forest-related reporting processes and organizations, with a view to pooling resources and streamlining reporting.

A total of 87 specialists from 45 countries and 17 international and regional organizations participated in this consultation.

The participants recognized that FRA 2005 is the most comprehensive assessment to date in terms of scope and the number of countries included. More than 800 specialists were involved over a period of four years – including 172 officially nominated national correspondents and their teams.

The experts noted that the increased country involvement and the network of national correspondents were key factors in the success of FRA 2005 and acknowledged the very substantial work by the national correspondents, who prepared the country reports. National correspondents underscored that the FRA reporting process offered an incentive to gather and analyse information on the forestry sector. They highlighted the importance of the country reports in assessing and monitoring forests at the national level and as input to the policy-making process.

The experts made a series of recommendations for the next Global Forest Resources Assessment, scheduled to take place in 2010:

- The topics covered by FRA 2005 were important and should be retained, with suggested changes to some of the existing tables.
- The use of sustainable forest management thematic elements as the reporting framework for FRA 2005 increased the relevance of the process and should be maintained for FRA 2010, with the addition of the seventh thematic element on the legal, policy and institutional framework.
- FRA 2010 should provide the forest-related information needed for the assessment of progress towards the 2010 biodiversity target of CBD.
- Country reporting should form the basis for FRA 2010, supplemented by special studies on specific issues and a remote sensing component providing complementary information on the spatial distribution of forests and on forest cover and land-use change dynamics at regional and global levels.
- The network of national correspondents should be maintained and strengthened, and regional networks should be supported by the countries and FAO.

The organizations participating in Kotka V (the Amazon Cooperation Treaty Organization, CBD, International Network for Bamboo and Rattan, ITTO, IUFRO, MCPFE, UNEP, UNEP-WCMC, UNFCCC, UNFF, World Agroforestry Centre [ICRAF] and World Bank) stressed the benefits they had received from FRA 2005. They confirmed their willingness to contribute information to future FRA work and to indicate their specific needs in order to promote streamlining of reporting. The meeting recommended that collaboration with forest-related organizations should be maintained and enhanced, with a view to pooling resources and expertise and reducing the reporting burden on countries.

It was further recommended that a longer-term strategy for FRA be developed. This should include: an analysis of the role and advantages of regional networks and regional reporting; the future reporting schedule and modality; and options for the further streamlining of reporting on forests at the international level. The next session of the FAO Committee on Forestry (COFO 2007) is expected to provide further guidance to FRA.

# Mountain development

**MOUNTAINS COVER** one-quarter of the earth's land surface and are home to more than 700 million people – most of them poor, isolated and marginalized. In the 15 years since UNCED, when mountains were recognized for the first time as being of global importance, mountain issues have gained increasing attention. Action in the field to improve the plight of mountain people and protect mountain environments is now widespread. The International Year of Mountains in 2002 provided a unique opportunity to devote attention to mountain issues and has led to increased support at many levels.

FAO/T. Hofer

Since the International Year of Mountains – 2002, efforts to improve the plight of remote mountain people and to protect mountain environments have become more widespread (Nepal)

Developments since the International Year of Mountains include the following.

- A new mountain convention came into force in the Carpathian region.
- Processes to enhance collaboration based on the Carpathian model are under way in the Balkans and Caucasus mountain ranges.
- Countries in the Hindu Kush Himalaya and the Andes have expressed interest in exploring mechanisms for transboundary cooperation.
- The CBD has developed a work programme on mountain biological diversity.
- The Mountain Partnership, launched at the 2002 World Summit on Sustainable Development, has expanded to include 130 members, including governments, international organizations, civil society groups, and private-sector members. The partnership facilitates networking, communications, livelihood improvements and sustainable rural development in mountain regions.
- Global efforts, including the Millennium Ecosystem Assessment and the Mountain Research Initiative, are creating greater awareness of mountain issues.

Overcoming poverty remains the greatest challenge. Mountain people are still among the world's poorest, and remoteness is often a barrier to development and to participation in the benefits of the global economy. Rapid advances in communications and information technology are helping to overcome the physical barriers faced by mountain communities.

# Payment for environmental services

**IT IS WELL KNOWN** that forests can produce a wide range of non-market benefits. Wider recognition of these benefits among policy-makers has been promoted in international and national policy debates on the management and use of forests.

Techniques for valuing non-market benefits have existed for several decades and have been refined to the point where they are now accepted in some (mostly developed) countries in public-sector project and policy appraisals. A more recent trend has been the development of mechanisms to reward forest owners for the production of non-market benefits, often called payments for environmental services.

There is increasing interest in this topic, as evidenced by the growing number of published studies (Figure 82).

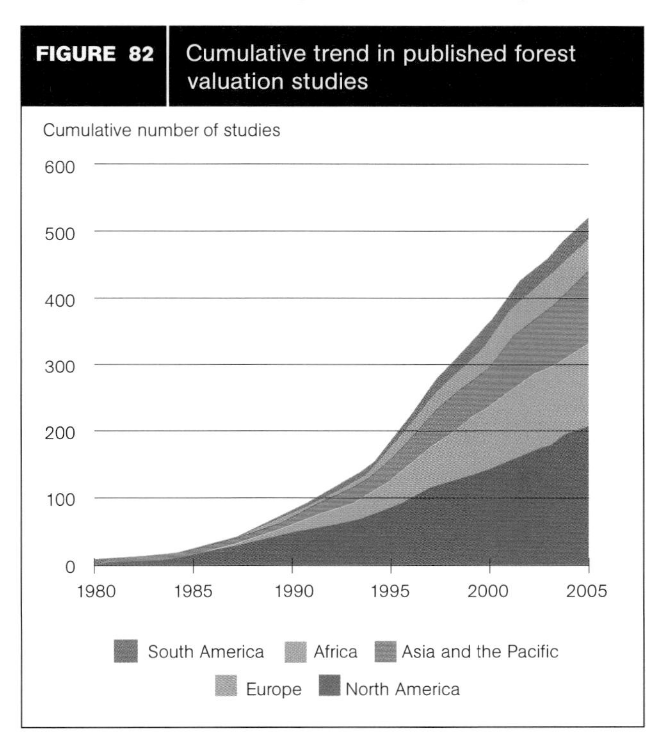


**FIGURE 82** Cumulative trend in published forest valuation studies

**SOURCE:** Derived from Envalue (www.environment.nsw.gov.au/envalue), EVRI (www.evri.ec.gc.ca) and FAO (www.fao.org/forestry/finance) databases on valuation of non-market benefits.

Most have been produced in developed countries: Australia and New Zealand (which account for a significant proportion of the Asia and the Pacific studies) and countries in Europe and North America. Many of these studies have examined recreation, amenity and environmental benefits (including the value of hunting, which has been a popular subject in North America). In developing countries, in contrast, most studies have examined the value of the subsistence use of forest products, rather than the broader social and environmental benefits.

Payments for environmental services include user fees (such as entry fees for recreation areas or permit fees for hunting) and artificial markets for other services from forests (e.g. payments for watershed protection activities). The latter have been a relatively recent innovation and, in many cases, these markets have been created and developed by governments.

Although only partial information is available, it is likely that payments for biodiversity services (especially user fees) are currently much larger than payments for carbon and watershed protection activities, as markets for the latter are much newer, and that formal arrangements for payments have been developed in only a few countries. A significant proportion of the market involves ad hoc payments or voluntary arrangements (e.g. investments in forestry for carbon sequestration external to the Kyoto Protocol).

Although interest in forest valuation and payment for environmental services remains high, many developing countries are unable to use these techniques owing to the high costs of data collection, analysis and the establishment of markets for such payments. In addition, many developing countries have difficulty collecting all the forest charges and taxes owed by producers of forest products. This suggests that addressing the latter problem should be a greater priority for immediate action than the developing of more sophisticated mechanisms such as these payments.

**PLANTED FORESTS CONTINUE** to expand, and their contribution to global wood production is approaching 50 percent of the total.

FAO gathered new information on planted forests in 2005 (FAO, 2006i), taking into account for the first time the planted component of semi-natural forests, which are neither strictly natural forests with minimal management nor forest plantations of introduced species with intensive management (Box 8). Semi-natural forests may be reforested by enrichment planting and/or seeding or through assisted natural regeneration and silvicultural treatments that enhance growth and yield.

The survey covered 38 selected countries representing 83 percent of the semi-natural forest area and 86 percent of the global forest plantation area.[1]

Asia leads the world in planted forests, followed by Europe (Figure 83). The areas of forests planted for productive purposes and of those planted for protective purposes are both steadily increasing (Figure 84). The trends are similar in all regions except Africa.

The top ten countries accounted for 81 percent of the 38 countries surveyed (Table 38). Of the planted forests in these countries, 73 percent were managed for productive purposes and 27 percent for protective purposes. However, it is obvious from the results that not all countries use the categories "productive" and "protective", given that it is unlikely that the planted forests in any country would actually fall 100 percent in either category, as reported by Brazil, Japan, Sweden and the United States of America.

Conifers dominate the productive planted forest category, accounting for 54 percent of area reported in 2005 (Figure 85), while broadleaves account for 39 percent. In the protective forests category, coniferous species account for 47 percent and broad-leaved species 31 percent (Figure 86).

The global area of semi-natural forests increased marginally from 251 million hectares in 1990 to 256 in 2000 and 261 in 2005. For the surveyed countries, semi-natural forests in 2005 comprised 53 percent planted and 47 percent assisted natural regeneration. These proportions are representative of semi-natural forests globally; however, proportions varied markedly among regions, subregions and selected countries.

The proportion of semi-natural forests managed through assisted natural regeneration decreased over the period 1990–2005, particularly in Europe and South and Southeast Asia (Figure 87). An exception was North America, where the proportion of assisted natural regeneration increased.

Globally, the proportion of semi-natural forests established through planting or seeding increased, especially in East Asia. A marginal decrease was registered in Africa.

TABLE 38

**Ten countries with largest area of planted forests, 2005 *(1 000 ha)***

| Country | Total | Productive | Protective |
|---|---|---|---|
| China | 71 326 | 54 102 | 17 224 |
| India | 30 028 | 17 134 | 12 894 |
| United States of America | 17 061 | 17 061 | 0 |
| Russian Federation | 16 963 | 11 888 | 5 075 |
| Japan | 10 321 | 0 | 10 321 |
| Sweden | 9 964 | 9 964 | 0 |
| Poland | 8 757 | 5 616 | 3 141 |
| Sudan | 6 619 | 5 677 | 943 |
| Brazil | 5 384 | 5 384 | 0 |
| Finland | 5 270 | 5 270 | 0 |
| **Total** | **181 693** | **132 095** | **49 597** |

**BOX 8** Planted forests in the continuum of forest characteristics

| Naturally regenerated forests | | | Planted forests | | | Trees outside forests |
|---|---|---|---|---|---|---|
| Primary | Modified natural | Semi-natural | | Plantations | | |
| | | Assisted natural regeneration | Planted component | Productive | Protective | |
| Forest of native species, where there are no clearly visible indications of human activities and the ecological processes are not significantly disturbed | Forest of naturally regenerated native species where there are clearly visible indications of human activities | Silvicultural practices by intensive management: • Weeding • Fertilizing • Thinning • Selective logging | Forest of native species, established through planting or seeding, intensively managed | Forest of introduced and/or native species established through planting or seeding mainly for production of wood or non-wood goods | Forest of introduced and/or native species, established through planting or seeding mainly for provision of services | Stands smaller than 0.5 ha; tree cover in agricultural land (agroforestry systems, home gardens, orchards); trees in urban environments; and scattered along roads and in landscapes |

[1] Figures for many other countries were obtained too late to be included in the present analysis.

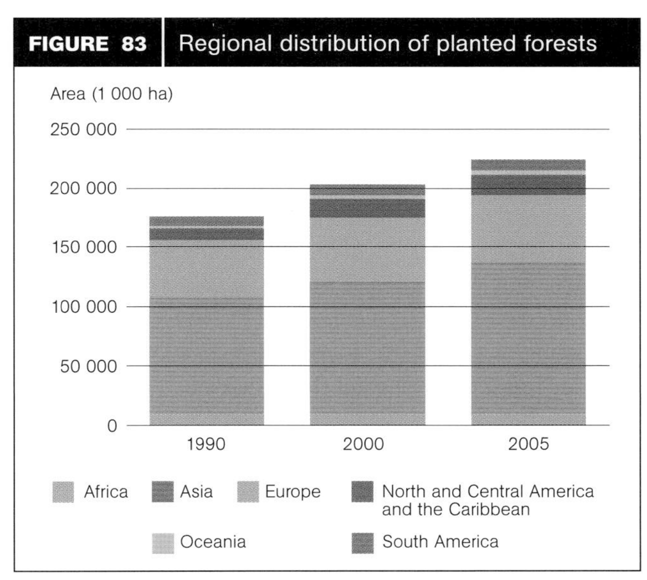


**FIGURE 83** Regional distribution of planted forests

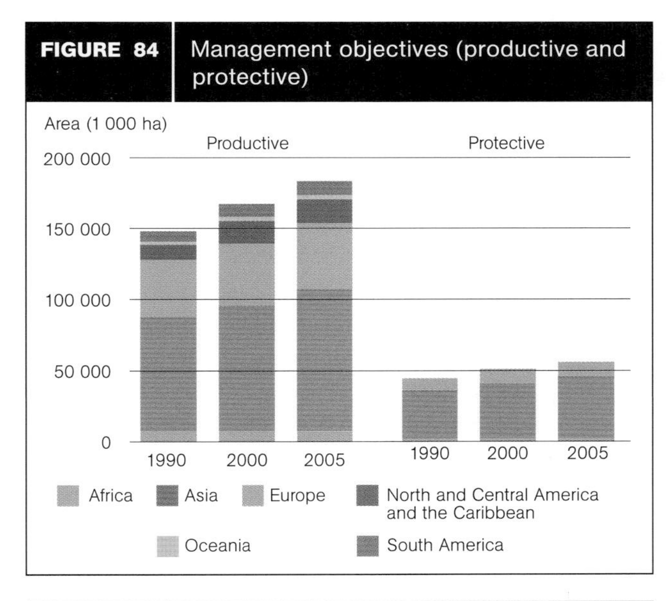


**FIGURE 84** Management objectives (productive and protective)

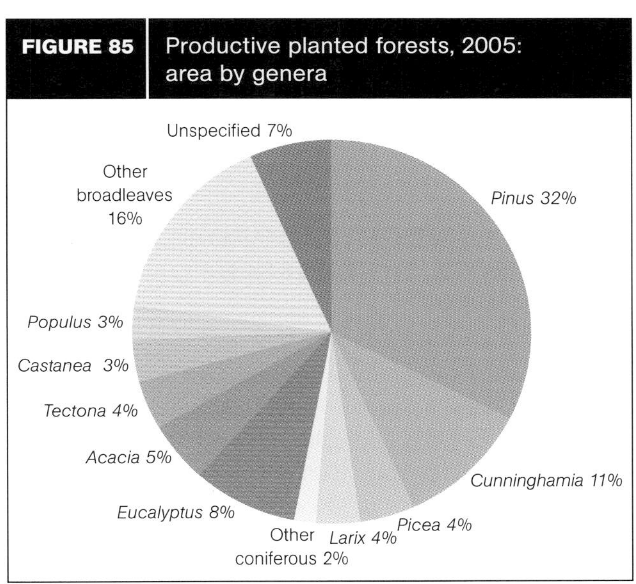


**FIGURE 85** Productive planted forests, 2005: area by genera

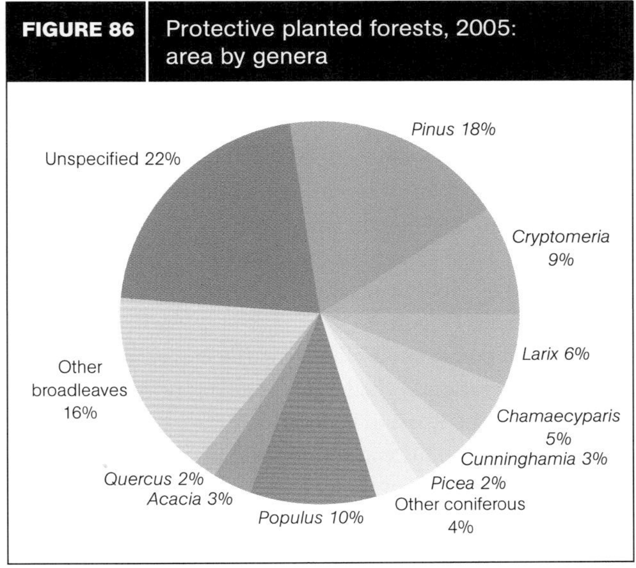


**FIGURE 86** Protective planted forests, 2005: area by genera

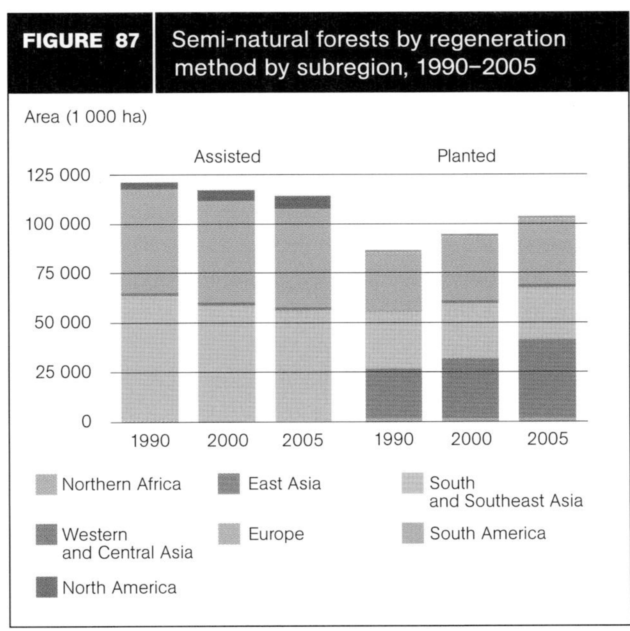


**FIGURE 87** Semi-natural forests by regeneration method by subregion, 1990–2005

**NOTE:** East and Southern Africa, West and Central Africa and Oceania reported 0.

# Trade in forest products

**PRODUCTION OF INDUSTRIAL** roundwood was 1.6 billion cubic metres in 2004, and about 7 percent of this was exported (about 120 million cubic metres). Thus 93 percent of industrial roundwood was processed domestically for domestic consumption or export.

Forest products trade reached a total value of US$327 billion in 2004 (Figure 88). This represents 3.7 percent of global trade value in all commodity products. Primary wood products accounted for 21 percent and primary paper products for 34 percent of the value of forest products trade. Secondary products (such as furniture or books) accounted for the remainder. The recent sharp increase in values is largely a result of the high appreciation of the euro to the United States dollar.

## TRENDS IN REGIONS AND PRODUCTS

On a global scale, the greater part of the forest products trade has taken place within Europe, within North America, and among Asia and the Pacific, Europe and North America. Europe is the largest exporting and importing region in the world. It imported a value of US$158 billion and exported US$184 billion of forest products in 2004, accounting for, respectively, 47 and 56 percent of global import and export value. These large shares are a result primarily of the high proportions of paper products and secondary processed products traded (Figure 89).

A recent conspicuous change is the emergence of the Russian Federation as the largest exporter of industrial roundwood. The Russian Federation exported 42 million cubic metres of industrial roundwood in 2004, accounting for 35 percent of global trade. East Asia and Europe are major importers of Russian timber.

Another interesting feature is that, from 2001, North America as a whole became a net importer of forest products (in terms of value). Moreover, this gap in net trade is widening every year because of rapidly growing imports to the United States of America from Asia, Europe and South America.

Wood-processing industries have developed in the past decade, especially in China, Eastern Europe and several developing countries. For example, in 2004 China became the largest importer of industrial roundwood, a major exporter and importer of wood-based panels, the second largest importer of paper and paperboard and the largest exporter of secondary processed wood products such as wooden furniture. Eastern European countries have become major exporters of sawnwood, wood-based panels and secondary processed wood products. Southeast Asia and Brazil have also developed their secondary wood-processing industries.

Foreign investment has played a critical role in the development of processing industries in rapidly growing regions, particularly in technology transfer, infrastructure development and improved access to global markets. Factors that have promoted foreign investment include low labour and production costs, government support to education and research, incentive policies for foreign investment and a growing domestic economy. Proximity to forest resources and major markets used to be a fundamental factor. However, as has been noted in China's exports, low production costs offset the higher transportation costs in reaching forest resources and global markets. The gap is widening between developing countries that are able to produce competitive products using foreign investment and those that are not able to do so.

Recent expansion of processing capacity in developing regions has shifted production bases on a global scale. A consequence is intensified competition, reflected in the downward trends in trade prices for major wood products. Faced with the rapidly increasing imports of wood products from China, the United States of America and the European Union have imposed anti-dumping duty on some Chinese products.

## DEVELOPMENT OF TRADE POLICIES TO PROMOTE SUSTAINABLE FOREST MANAGEMENT

### Public procurement policy

Several countries (including Belgium, Denmark, France, Germany, Japan, the Netherlands, New Zealand, Sweden and the United Kingdom) have developed or are developing public procurement policies to promote the use of legally or sustainably produced products, as have several local governments in Europe and the United States of America. Verification of legality is a basic requirement in these schemes, although differences are seen with regard to criteria, sources and coverage of products, and methods of verification.

### Private-sector initiatives

"Green building" initiatives in a number of countries promote construction practices that use sustainably produced products. Examples include the Leadership in Energy and Environmental Design Green Building Rating System by the United States Green Building Council, which gives points for the use of certified wood. Similar initiatives have started in Canada and Europe.

Recently, some large European paper companies have begun incorporating chain-of-custody verification and certification of forest management into their investment projects in developing countries. At their second global meeting in June 2006 in Rome, Italy, the chief executive officers of 54 international forest industry companies signed a Commitment to Global Sustainability.

### Phytosanitary measures

In an effort to control the spread of invasive pests, in 2002 the Interim Commission on Phytosanitary Measures of the IPPC adopted International Standard for Phytosanitary Measures No. 15 (ISPM 15) for treating wood packaging material in international trade. As of January 2006, the European Union and more than 20 countries have implemented or are developing national standards in accordance with ISPM 15, including major exporters and importers of industrial commodities that use wood packaging material.

**FIGURE 88** Regional trends in forest products trade

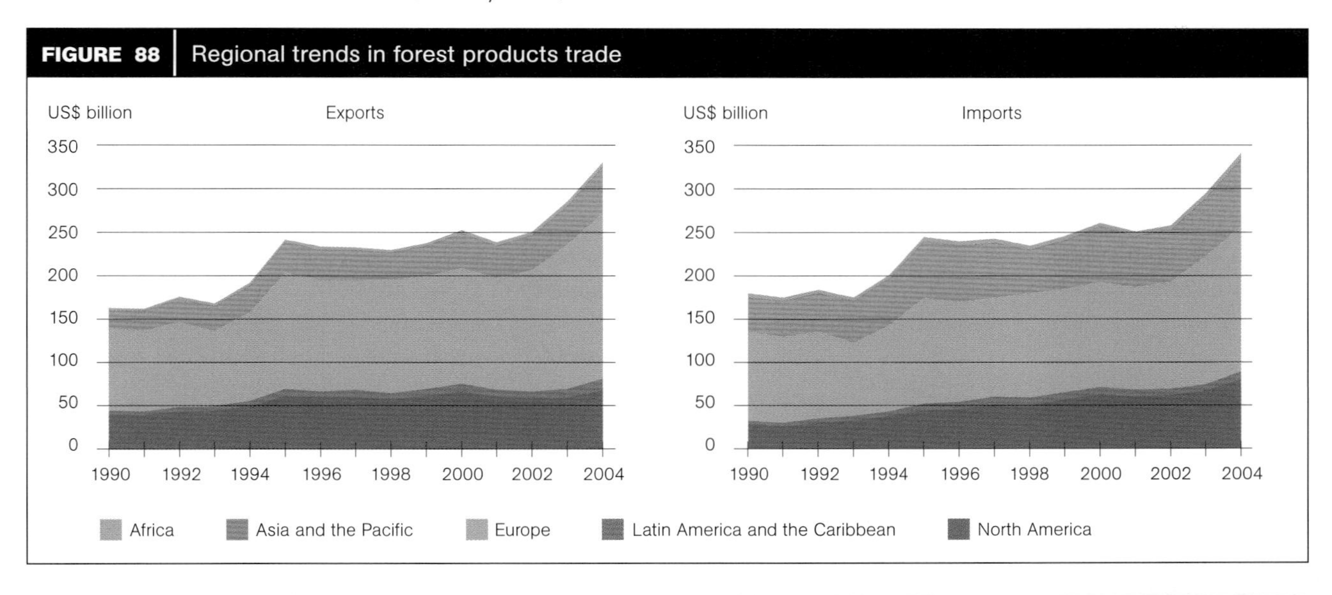


**FIGURE 89** Global exports of forest products, 1990–2004

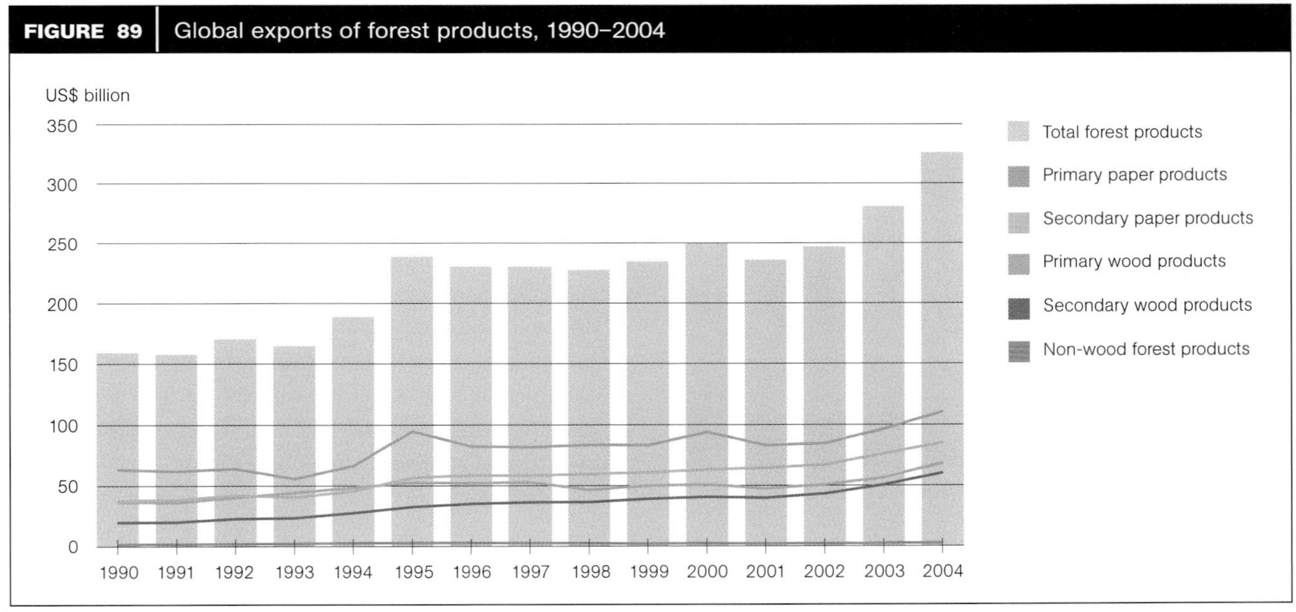


**NOTE:** Primary wood products include roundwood, sawnwood, wood-based panels and wood chips. Secondary wood products include wooden furniture, builders' joinery and carpentry. Primary paper products include pulp, paper and paperboard. Secondary paper products include packaging cartons, boxes and printed articles, including books and newspapers.

**SOURCES:** FAO, 2006b; United Nations, 2006.

**THE URBANIZATION** of society continues to increase, presenting challenges and opportunities for forestry. As urban areas expand, nearby trees and forest resources are usually lost or degraded. At the same time, there is a growing awareness across the globe of the importance of urban green spaces to the quality of the urban environment and urban life.

Cities in developing countries face specific problems in the supply of essential products to city dwellers, including food, wood-based energy and clean water. Urban residents fight pollution and search for the recreational and leisure functions provided by green areas. The development of human settlements is often spontaneous and uncontrolled, particularly in situations of conflict or natural disaster.

Urbanization has a heavy impact on the natural resource base, including the use of forests and trees for basic wood products and fuelwood. This can result in watershed degradation and soil erosion in the rural areas surrounding cities. On the other hand, the rural poor can benefit from the income generated by the production of wood, fuelwood, NWFPs and foodstuffs if there are rights to tree and forest resources and equitable access.

Urban greening in Central and Western Asia: a community tree-planting day in Kabul, Afghanistan (2006)

The challenge of urban forestry has been increasingly addressed in international venues, for example the IUFRO World Congress 2005; the eighth and ninth sessions of the European Forum on Urban Forestry in 2005 and 2006; and the second and third sessions of the Urban World Forum in 2004 and 2006. Much of the participation in these events, however, has been from developed countries. A challenge is to make it possible for the cities of the developing world to benefit from the lessons learned in the developed world.

The Asia–Europe Meeting (ASEM) is an informal process of dialogue and cooperation that brings together the member countries of the European Union with 13 Asian countries. This process has sponsored two symposia on urban forestry. The first, held in China in 2004, resulted in a set of goals, priorities and follow-up actions for cooperation in urban forestry among member countries. The second, held in Denmark in June 2006, focused on urban forestry for human health and well-being.

In April 2006, FAO brought together representatives from five countries in Central and Western Asia to consider ways in which urban and peri-urban forestry could help alleviate poverty. Cities in these countries have similar problems: water quality, degradation of forest-based resources, and poverty. Examples were presented of good practices in the planning, management and use of urban greening in the region that have successfully contributed to urban livelihoods and quality of life. Workshop participants recommended that the social, cultural, economic and environmental benefits of urban greening need to be assessed and marketed and raised on the agenda of municipal and governmental policy-makers.

**MANY NATIONAL** forest policies and international commitments stress that sustainable management of forests can be promoted through voluntary tools such as management guidelines, codes of best practices, criteria and indicators and standards for certification (Table 39). At the international level, such tools are developed through partners working together towards a common goal and defining shared principles and mechanisms. These voluntary instruments enhance knowledge-sharing and provide the means to conceptualize sustainable forest management, implement it and evaluate progress towards achieving it.

## NATIONAL AND LOCAL GOVERNMENTAL TOOLS

Sustainable forest management requires a solid legal and policy foundation at the national level – or at the subnational level in countries where the responsibility for managing forests has been devolved to this level.

In the 15 years since UNCED, a majority of the world's countries have established or updated their national forestry laws and policies and are moving towards integrated approaches that balance environmental, economic and social aspects of forest management. In many countries, significant steps have been taken to devolve forest management to local levels and to involve local people in decision-making.

Some of the most innovative national and local policies are originating in tropical countries, where the battle to slow deforestation is fought on a daily basis. A vivid example is Costa Rica, the only country in Latin America that has succeeded in reversing the downward trend in forest area. Costa Rica is implementing many different tools to promote sustainable forest management, including tax incentives and payment for environmental services.

## INTERGOVERNMENTAL INITIATIVES

### Non-legally binding instruments for all types of forests

To strengthen political commitment and action to implement sustainable management of all types of forests and to achieve the global objectives on forests, UNFF has agreed to adopt a non-legally binding instrument on all types of forests by 2007 (ECOSOC, 2006). It remains to be seen whether this instrument will be more or less effective than the existing Forest Principles adopted at UNCED in 1992.

### Criteria and indicators for sustainable forest management

Criteria and indicators are used to monitor, assess and report on progress towards sustainable forest management. Most of the member countries of ITTO, MCPFE and the Montreal Process produce periodic reports on status and trends in forestry using the C&I framework. It is also applied in national forest programmes, certification and in communicating progress to policy-makers and the public.

CPF members and many national governments continue to promote the implementation of C&I as a framework for reporting and as a tool to support improved forest management practices. In 2004, four Central Asian countries joined the C&I process. Member countries of the Lepaterique Process of Central America recently reaffirmed their commitment to continue using C&I for reporting on progress towards sustainable forest management. The eight Amazon countries of the Tarapoto Process, operating under the umbrella of the Organización del Tratado de Cooperación Amazónica, validated 15 priority indicators in 2006.

ITTO used the C&I reports of its producer member countries as a basis for the *Status of tropical forest management 2005*. IUFRO promotes the use of C&I through training and through involving scientific and academic societies in the C&I process. CIFOR supports countries in improving their C&I for community-based and participatory forest management.

### Forest law enforcement and governance

In recent years, a number of countries have actively promoted improvements in forest law enforcement and governance as a key component of national and international efforts to achieve sustainable forest management. Most of the initiatives have been regional in nature, for example in Central Africa, East Asia and Europe. International organizations actively involved in supporting the initiative include FAO, ITTO and the World

TABLE 39
**Examples of tools to promote sustainable forest management**

| | Voluntary tools | Legally binding instruments |
|---|---|---|
| National | National forest programmes | National laws and regulations |
| Regional | C&I processes | Regional conventions |
| Global/international | UNCED Forest Principles | International Tropical Timber Agreement |

Bank. The Group of Eight (G8) countries have also played a catalytic role.

## NON-GOVERNMENTAL INITIATIVES

Certification schemes are market-based tools for sustainable forest management. The logic of certification is simple: if consumers prefer products from forests that are certified as being managed on a sustainable basis, or if they are willing to pay a higher price for certified products, then forest producers will have an incentive to adopt sustainable forest management practices.

The area of certified forests has expanded rapidly in recent years. The total area of certified forests is approaching 20 percent of the world's production forests (as defined by FRA 2005), although this is only 7 percent of the world's forest area. The majority of certified forests continue to be found in developed countries, where forest area was already stable or increasing before certification arrived on the scene. The challenge for certification is the need to expand the process to tropical forests.

## TOOLS DEVELOPED BY INTERNATIONAL AGENCIES, INCLUDING MULTISTAKEHOLDER INITIATIVES

The development of voluntary guidelines is a key part of many international agencies' work, including that of FAO and ITTO (Box 9). They vary in scope and level – from detailed operational practices to broader policy guidelines, and from regional to global. The more successful of these initiatives have one thing in common: they were developed by a broad spectrum of stakeholders representing government, the private sector and civil society.

Voluntary guidelines on planted forests and fire management are currently being developed through a broad consultative process involving technical experts representing different sectors and regions, the regional forestry commissions and COFO.

Once the voluntary guidelines are established, regional training workshops will be conducted to strengthen countries' capacity to translate the principles into policies and practices. Following the example of the *FAO model code of forest harvesting practice* (FAO, 1996), regions and countries can adapt and apply the new guidelines to their local situations and conditions.

## USE AND USEFULNESS OF VOLUNTARY TOOLS

Voluntary, non-legally binding instruments provide guiding principles on forest use. At their best, they build upon international agreements and commitments, most importantly the Rio Forest Principles, but also international labour and trade agreements, including the International Tropical Timber Agreement (ITTA).

**BOX 9** Examples of forest-related voluntary guidelines

Since 1990 countries have worked to develop and implement criteria and indicators for sustainable forest management through regional and international processes, pioneered by ITTO, that now cover 155 countries.

ITTO developed *Guidelines on the conservation of biological diversity in tropical production forests* (ITTO, 1993) to optimize the contribution of timber-producing tropical forests to the conservation of biological diversity.

The *FAO model code of forest harvesting practice* (FAO, 1996) was compiled to highlight the wide range of environmentally sound harvesting practices available and enable policy-makers to develop national, regional or local codes of practice to serve particular needs. Subsequently, regional codes were agreed in Asia and the Pacific in 1999 (FAO, 1999) and West and Central Africa (FAO, 2005c). National-level codes have been adopted or are under preparation in several countries in Southeast Asia.

*Governance principles for concessions and contracts in public forests* (FAO, 2001b) compiles critical factors in balancing and safeguarding the public and private interest in forest management and identifies new approaches to contractual arrangements in the provision of goods and services from public forests.

ITTO, in collaboration with partners, developed *Guidelines for the restoration, management and rehabilitation of degraded and secondary tropical forests* (ITTO, 2002), which is part of ITTO's series of internationally agreed policy documents for achieving the conservation, sustainable management, use and trade of tropical forest resources.

The Confederation of European Paper Industries developed *Legal logging: code of conduct for the paper industry* (CEPI, 2005) to combat illegal logging.

*Best practices for improving law compliance in the forest sector* (FAO/ITTO, 2005) distils the available knowledge that decision-makers could follow in reducing illegal operations in the forest sector.

The implementation of voluntary tools varies among regions and countries. For example, the *Regional code of practice for reduced-impact forest harvesting in tropical moist forests of West and Central Africa* (FAO, 2005c) is not yet well integrated at the national level owing to a lack of resources for training, while the *Code of practice for forest harvesting in Asia–Pacific* (FAO, 1999) is being implemented through national codes, with support from the Asia–Pacific Forestry Commission and bilateral donors.

Tremendous progress has been made in the conceptual development of criteria and indicators. However, C&I implementation lags behind in most developing countries, owing to deficiencies in data collection, analysis and storage, and weak institutional capacity, which affect the proper use and implementation of C&I. Their implementation is much farther along in countries that have significant financial resources.

The best results in developing countries have been achieved by linking C&I with national forest assessments and inventories and national forest programmes.

The area of certified forest has increased in developed countries and is helping to improve management practices, yet the original objective of combating deforestation in the tropics has not been widely realized. The problems are fairly straightforward: most deforestation in the tropics is caused by conversion to other land uses rather than by logging; certification is not cheap; and its successful implementation requires a solid institutional and governance platform. Moreover, it is yet to be seen whether consumers are willing to pay more for certified products on a large scale.

International agencies organize national and regional workshops for training and exchange of experiences to boost the use of the tools discussed. The challenge is to provide adequate support to capacity-building so that countries can make the best use of voluntary instruments. An emerging challenge is to avoid overlaps between different instruments and to integrate them through national policy and monitoring frameworks.

**RECENT STUDIES** have emphasized the complexity of the relationship between forests and water, including the myths that more trees are always "good" and that deforestation is always "bad".

*Floods in Bangladesh: history, dynamics and rethinking the role of the Himalayas* (Messerli and Hofer, 2006) concludes that there is no evidence for a direct and causative link between Himalayan deforestation and floods in Bangladesh. The impact of forest cover on flooding is a question of scale: the effects of forest clearing on flood flows and sediment transport are immediate and strong in small mountain watersheds, while in large river basins, natural processes dominate.

*Forests and floods: drowning in fiction or thriving on facts?* (FAO/CIFOR, 2005) concludes that "... direct links between deforestation and floods are far from certain" and that "All floods cannot and should not be completely prevented – flooding is important for maintaining biodiversity, fish stocks and fertility of floodplain soils".

*From the mountain to the tap: how land use and water management can work for the rural poor* (DFID, 2005) drew several conclusions that startled some foresters, including the provocative statement that "Trees on the whole are not a good thing in dry areas if you want to manage water resources". Many trees, especially fast-growing species such as pines and eucalyptus, suck more water from the ground than other crops. The water transpires from the leaves, and the trees contribute to drying out the land. The report identifies ten policy lessons, emphasizing the importance of policy instruments and market mechanisms that benefit the poor and give due attention to livelihood benefits, not just water allocation.

At the World Water Council's fourth World Water Forum in Mexico City, Mexico, in March 2006, ministers and scientists debated a World Bank paper, *Water, growth and development* (World Bank, 2006b), which argued that investments in water infrastructure will automatically lead to development. Several participants argued that one single approach will not work throughout the developing world.

A review by FAO and international partners (FAO, 2006k) – including the European Observatory of Mountain Forests (EOMF), the International Centre for Integrated Mountain Development (ICIMOD), REDLACH and ICRAF – recommended new and innovative approaches to watershed management, including the following:

- a move from participatory to collaborative approaches to managing watersheds (Table 40);
- more attention to institutional aspects;
- flexible programme design;
- a long-term approach to planning and financing for watershed management.

A thematic study on forests and water carried out in the framework of FRA 2005 (FAO, 2006l) identifies categories of forest ecosystems that require special attention based on their hydrological relevance:

- mountain cloud forests;
- swamp forests;
- forests on saline-susceptible soils;
- forests on sites with landslip risk;
- riparian forests;
- municipal water-supply forests;
- avalanche protection forests.

TABLE 40
**Participatory and collaborative watershed management compared**

| Participatory watershed management | Collaborative watershed management |
|---|---|
| Focusing on communities and people and targeting primarily grassroots social actors (households, small communities) | Focusing on civil society and targeting a variety of social and institutional actors, including local governments, line agencies, unions, enterprises and other civil society organizations, as well as technical experts and policy-makers |
| Assuming that sound natural resource management is a public interest shared by all social actors | Recognizing that stakeholders bear particular (and sometimes contrasting) interests in natural resources that need to be accommodated |
| Seeking (or claiming) to make decisions through a bottom-up process, by which grassroots aspirations are progressively refined and then turned into operational statements and action | Seeking decision-making that merges stakeholders' aspirations and interests with technical experts' recommendations and policy guidelines through a continued two-way (bottom-up and top-down) negotiation process |
| Centring on the watershed management programme, with local government expected to assist as a side supporter | Centring on the local governance process, with the watershed programme acting as facilitator and supporter |
| Aiming at general consensus, presuming that conflict can be solved through dialogue and participation | Aiming at managing natural resource conflicts from the awareness that often dialogue and participation can mitigate (partially and temporarily) conflicts, but not fix them structurally |

**AMONG THE MULTIPLE** threats to wildlife, two of the most immediate and direct are unsustainable hunting and trading in wildlife and wildlife products, and human–wildlife conflict.

In many parts of Africa, commercial trade in bushmeat for consumption is probably the single most important cause of the decline of wildlife populations, ranging from insects, birds and turtles to primates, antelopes, elephants and hippopotamuses. It was estimated that in the Congo Basin, alone, the annual offtake of bushmeat is about 5 million tonnes (Fa, Peres and Meeuwig, 2002), but a recent, detailed study of bushmeat offtake in the moist forests of Cameroon and Nigeria (Fa *et al.*, 2006), which documented an average offtake of 346 kilograms per square kilometre, suggests a much lower offtake of up to 1 million tonnes for the Congo Basin. However, this lower estimate gives little cause for comfort, because it is still far in excess of a sustainable level, given the inherently low production of animal biomass in tropical forests.

Meat from wild animals is not an African issue only (Table 41). The meat from freshwater turtles is consumed in huge volumes in East Asia, despite the fact that three-quarters of the 90 species found in Asia are considered threatened, and 18 of those critically endangered (IUCN, 2005).

There are success stories of the revival of overexploited wild animal populations. In 1969, all 23 species of crocodilians were threatened or had declining populations. Today, one-third of crocodilians can sustain a regulated commercial harvest, and only four species are critically endangered. In many cases, well-managed, CITES-approved ranching programmes produce sustainably harvested hides for the international market, garnering the support of industry and governments, while helping supplant illicit trade. Similar programmes in regulating the trade in wool products from South America's vicuna have resulted in similar successes. By the 1960s, vicuna populations had been reduced to 5 000 animals, less than 1 percent of historical populations, but conservation and management have restored their numbers to 160 000. Today, the illegal global trade in wildlife is second only to narcotics and is valued at almost US$5 billion (Wildlife First, 2006).

Because of human population growth, the accompanying growth of human settlements and the consequent reduction of wildlife habitat, conflicts between humans and wildlife are occurring more and more frequently around the world. In Africa, where many people depend directly on natural resources for their livelihoods, wildlife species such as crocodiles, elephants, hippopotamuses and lions raid crops, injure or kill livestock, invade human settlements and cause damage to personal belongings, and can even injure and kill people. As a result, local people are increasingly hostile to wildlife, and local communities do not cooperate with conservation authorities. The result is increased instances of poaching and other illegal activities.

The causes of human–wildlife conflicts will not be eliminated in the near future, and it can be expected that conflict will only increase in frequency and intensity. There is, therefore, an urgent need to find ways to manage human–wildlife conflict. A range of approaches are being tried, including natural and artificial barriers, such as suspending chilli-pepper-impregnated cloths on ropes surrounding agricultural fields, a technique used successfully in an FAO project in Ghana to deter elephants from raiding crops. At present, the most reasonable approach to managing human–wildlife conflict is to implement short-term mitigation strategies jointly with long-term preventive measures.

A poignant indicator that something is wrong in the human–wildlife relationship is the phenomenon described as "elephant breakdown" (Bradshaw *et al.*, 2005): "Elephant society in Africa has been decimated by mass deaths and social breakdown from poaching, culls and habitat loss.... Wild elephants are displaying symptoms associated with human post-traumatic stress disorder (PTSD): abnormal startle response, depression, unpredictable, asocial behaviour and hyperaggression." This phenomenon has recently been given as an explanation for the killing of rhinoceroses by hyperaggressive, young male African elephants. New conservation strategies are required to preserve elephant social systems and promote normal social patterns.

A challenge for policy-makers is to balance conservation of wildlife resources with the livelihood requirements of local populations, in all regions.

TABLE 41
**Decline of selected animal populations**

| Species | Initial population | Year | Current population | Decline (%) |
|---|---|---|---|---|
| Bonobo (pygmy chimpanzee) | 100 000 | 1984 | 5 000 | 95.0 |
| Asian elephant | 200 000 | 1900 | 40 000 | 80.0 |
| African elephant | 10 000 000 | 1900 | 500 000 | 95.0 |
| Tibetan antelope | 1 000 000 | 1900 | 75 000 | 92.5 |

**WOOD IS INCREASINGLY** used for energy. High fossil fuel prices together with new energy and environmental policies are making woodfuel an essential ingredient of energy policy in both developed and developing countries. In developed countries, it is likely that the use of wood for energy will continue to increase if fossil fuel prices continue to rise. More generally, the use of biofuels, including those based on wood and on agricultural products, will likely continue to increase, including their use for motor vehicles. In developing countries, wood is already the primary source of energy for heating and cooking: in Africa, almost 90 percent of all wood removals are used for energy. With ever higher fuel prices, there will be even more pressure on forests and trees outside forests to provide energy in the poorest countries.

Traditionally, the main sources of wood for fuel are woody residues and wastes derived from timber industries (sawmills, particle board and pulp mills). In poor rural areas of developing countries, fuelwood is usually obtained directly by felling trees or collecting fallen wood. Recently, recovered woody biomass and residues from logging operations have also become important supply sources.

In 2003, renewable energy accounted for 13.3 percent of the world's total primary energy supply (Figure 90). Biofuels amounted to almost 80 percent of total renewable energy. They supply more energy than nuclear sources, and about four times as much as hydropower, wind, solar and geothermal energy combined. About 75 percent of biofuels are derived from fuelwood, charcoal and black liquor (a by-product of pulp and paper-making).

Most biofuels are used for residential cooking and heating, mainly in Africa, Asia and Latin America. For example, almost 90 percent of the wood removals in Africa are used for fuel. In Organisation for Economic Co-operation and Development (OECD) countries such as Austria, Finland, Germany and Sweden, biofuels are increasingly used for the production of electricity, attracting huge investments in wood-energy industries. There is a growing market for forest by-products as raw materials for energy. Sawmills and pulp and paper industries benefit by becoming energy producers.

Outlook studies by the International Energy Agency indicate that renewable energy sources will continue to increase their market shares in the energy mix (IEA, 2005). While heating and cooking will remain the principal uses for fuelwood and charcoal in developing countries, the use of solid biofuels for the production of electricity is expected to triple by 2030 (Figure 91).

Although most current woodfuels are derived from by-products (residues and wastes), in the future more will be derived directly from forests and tree plantations. The positive and negative implications of increased use of wood as fuel will depend on the rationality of future energy, environmental, forestry and industrial policies, including the role of incentives and taxes for the promotion of wood as fuel.

International trade in woodfuels is expected to increase in some regions, including Central and South America. Woodfuel production and export could become key ingredients for the development and expansion of forest

**FIGURE 90** Fuel shares of world total primary energy supply,[a] 2003

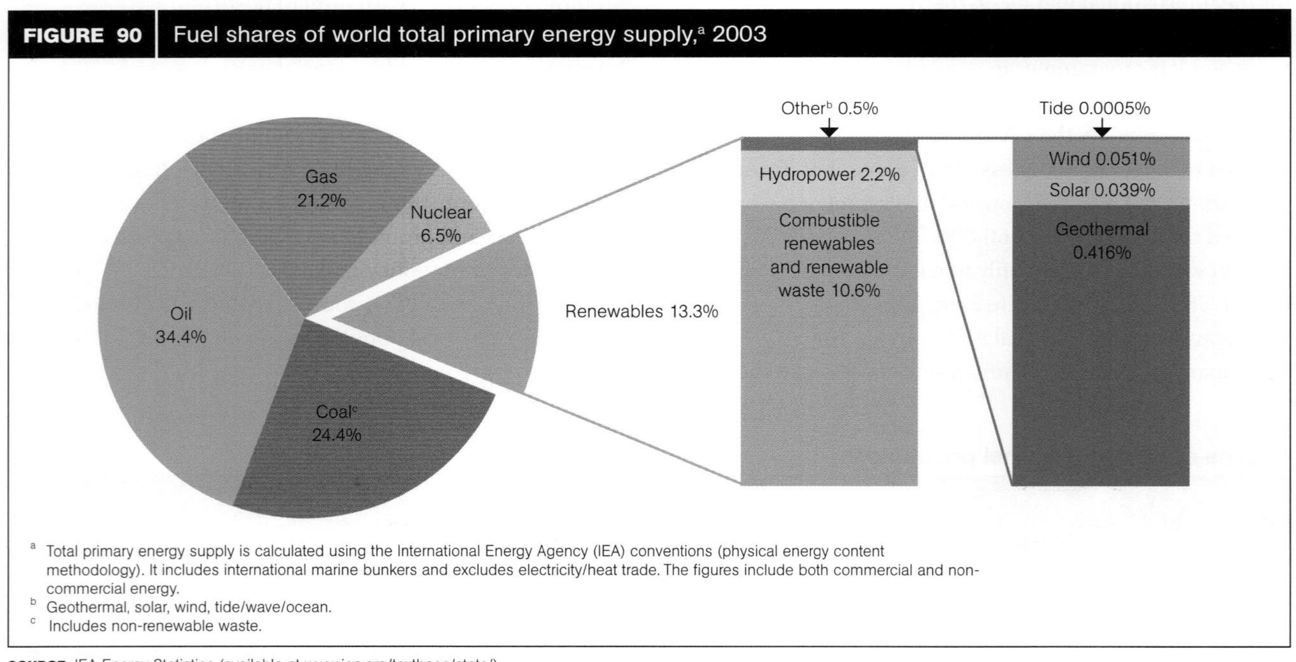


[a] Total primary energy supply is calculated using the International Energy Agency (IEA) conventions (physical energy content methodology). It includes international marine bunkers and excludes electricity/heat trade. The figures include both commercial and non-commercial energy.
[b] Geothermal, solar, wind, tide/wave/ocean.
[c] Includes non-renewable waste.

**SOURCE:** IEA Energy Statistics (available at www.iea.org/textbase/stats/).

activities, although it is not likely that this trend will have a direct impact on poverty. However, these activities may contribute to deforestation and forest degradation if policies are not implemented to avoid negative impacts.

As the demand for woody biomass for energy increases, structural changes in the energy sector will have positive and negative implications for wood industries. Wood energy could become a motor for the development and expansion of forestry activities. Progressive policies are required to ensure that these changes help alleviate poverty in developing countries.

Much of the research in this area has been undertaken by energy and forestry organizations in isolation from each other. This is clearly a field in need of more effective sharing of knowledge across traditional sectors.

**FIGURE 91** Supply of renewables by energy source

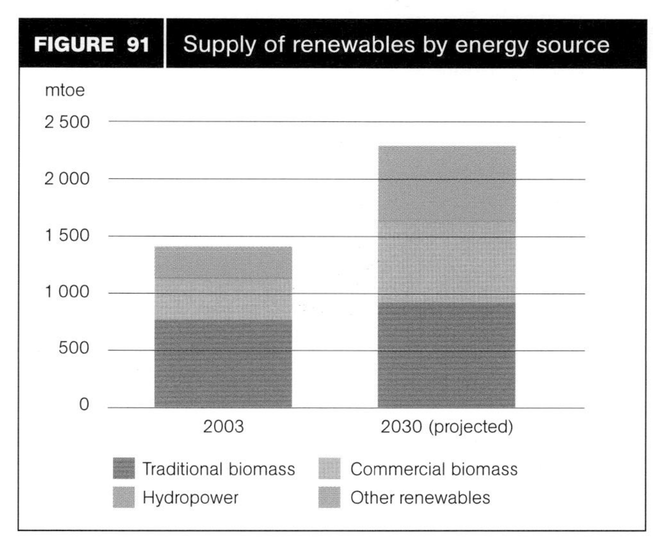


**NOTE**: Mtoe = million tonnes of oil equivalent.
**SOURCE:** IEA Energy Statistics (available at www.iea.org/textbase/stats/).

# Annex

TABLE 1
## Basic data on countries and areas

| Country/area | Land area[a] | Population 2004 | | | | GDP 2004 | |
|---|---|---|---|---|---|---|---|
| | | Total | Density | Annual growth rate | Rural | Per capita | Annual growth rate |
| | *(1 000 ha)* | *(1 000)* | *(Population/ km²)* | *(%)* | *(% of total)* | *(US$)* | *(%)* |
| Burundi | 2 568 | 7 343 | 285.9 | 1.9 | 89.7 | 104 | 5.5 |
| Cameroon | 46 540 | 16 400 | 35.2 | 1.9 | 47.9 | 651 | 4.8 |
| Central African Republic | 62 298 | 3 947 | 6.3 | 1.7 | 56.8 | 232 | 0.9 |
| Chad | 125 920 | 8 823 | 7.0 | 2.8 | 74.6 | 277 | 31.0 |
| Congo | 34 150 | 3 855 | 11.3 | 2.6 | 46.1 | 956 | 4.0 |
| Democratic Republic of the Congo | 226 705 | 54 775 | 24.2 | 3.0 | 67.7 | 89 | 6.3 |
| Equatorial Guinea | 2 805 | 506 | 18.0 | 2.4 | 51.0 | 3 989 | 10.0 |
| Gabon | 25 767 | 1 374 | 5.3 | 2.2 | 15.6 | 3 859 | 2.0 |
| Rwanda | 2 467 | 8 412 | 341.0 | 2.8 | 79.9 | 263 | 3.7 |
| Saint Helena | 31 | 7 | 24.1 | – | – | – | – |
| Sao Tome and Principe | 96 | 161 | 167.3 | 2.0 | 62.1 | 342 | 4.5 |
| **Total Central Africa** | **529 347** | **105 603** | | | | | |
| | | | | | | | |
| British Indian Ocean Territory | 8 | 1 | 15.0 | – | – | – | – |
| Comoros | 186 | 614 | 275.6 | 2.4 | 64.4 | 361 | 1.9 |
| Djibouti | 2 318 | 716 | 30.9 | 1.4 | 15.9 | 861 | 3.0 |
| Eritrea | 10 100 | 4 477 | 44.3 | 2.0 | 79.6 | 163 | 1.8 |
| Ethiopia | 100 000 | 69 961 | 70.0 | 1.9 | 84.1 | 113 | 13.4 |
| Kenya | 56 914 | 32 447 | 57.0 | 1.7 | 59.5 | 343 | 2.1 |
| Madagascar | 58 154 | 17 332 | 29.8 | 2.6 | 73.2 | 239 | 5.3 |
| Mauritius | 203 | 1 234 | 608.0 | 1.0 | 56.5 | 4 289 | 4.2 |
| Mayotte | 37 | 172 | 459.9 | – | – | – | – |
| Réunion | 250 | 777 | 310.8 | – | – | – | – |
| Seychelles | 45 | 85 | 188.2 | 1.3 | 49.9 | 6 573 | -2.0 |
| Somalia | 62 734 | 9 938 | 15.8 | 3.2 | 64.6 | – | – |
| Uganda | 19 710 | 25 920 | 131.5 | 2.5 | 87.7 | 285 | 5.7 |
| United Republic of Tanzania | 88 359 | 36 571 | 41.4 | 1.9 | 63.6 | 322 | 6.3 |
| **Total East Africa** | **399 018** | **200 245** | | | | | |
| | | | | | | | |
| Algeria | 238 174 | 32 373 | 13.6 | 1.7 | 40.6 | 1 981 | 5.2 |
| Egypt | 99 545 | 68 738 | 69.1 | 1.7 | 57.8 | 1 663 | 4.3 |
| Libyan Arab Jamahiriya | 175 954 | 5 674 | 3.2 | 2.1 | 13.4 | 7 483 | 4.5 |
| Mauritania | 102 522 | 2 906 | 2.8 | 2.0 | 37.0 | 396 | 6.6 |
| Morocco | 44 630 | 30 586 | 68.5 | 1.6 | 41.9 | 1 302 | 3.5 |
| Sudan | 237 600 | 34 356 | 14.5 | 2.4 | 60.2 | 448 | 6.0 |
| Tunisia | 15 536 | 10 012 | 64.4 | 1.2 | 35.9 | 2 315 | 5.8 |
| Western Sahara | 26 600 | 274 | 1.0 | – | – | – | – |
| **Total Northern Africa** | **940 561** | **184 919** | | | | | |

– = not available
0 = true zero
**NOTE:** The regional breakdown reflects geographic rather than economic or political groupings.

| Country/area | Land area[a] | Population 2004 | | | | GDP 2004 | |
|---|---|---|---|---|---|---|---|
| | | Total | Density | Annual growth rate | Rural | Per capita | Annual growth rate |
| | (1 000 ha) | (1 000) | (Population/ km²) | (%) | (% of total) | (US$) | (%) |
| Angola | 124 670 | 13 963 | 11.2 | 3.2 | 63.6 | 887 | 11.2 |
| Botswana | 56 673 | 1 727 | 3.1 | 0.3 | 48.0 | 3 684 | 4.6 |
| Lesotho | 3 035 | 1 809 | 59.6 | 0.9 | 81.9 | 540 | 3.1 |
| Malawi | 9 408 | 11 182 | 118.9 | 2.0 | 83.3 | 165 | 3.8 |
| Mozambique | 78 409 | 19 129 | 24.4 | 1.8 | 63.2 | 270 | 7.8 |
| Namibia | 82 329 | 2 033 | 2.5 | 0.9 | 67.0 | 1 905 | 4.2 |
| South Africa | 121 447 | 45 584 | 37.5 | – | 42.6 | 3 307 | 3.7 |
| Swaziland | 1 720 | 1 120 | 65.1 | 1.3 | 76.3 | 1 356 | 2.1 |
| Zambia | 74 339 | 10 547 | 14.2 | 1.4 | 63.8 | 366 | 4.7 |
| Zimbabwe | 38 685 | 13 151 | 34.0 | 0.4 | 64.6 | – | – |
| **Total Southern Africa** | **590 715** | **120 245** | | | | | |
| | | | | | | | |
| Benin | 11 062 | 6 890 | 62.3 | 2.5 | 54.7 | 389 | 2.7 |
| Burkina Faso | 27 360 | 12 387 | 45.3 | 2.3 | 81.8 | 257 | 3.9 |
| Cape Verde | 403 | 481 | 119.4 | 2.5 | 43.3 | 1 328 | 5.5 |
| Côte d'Ivoire | 31 800 | 17 142 | 53.9 | 1.8 | 54.6 | 583 | −2.3 |
| Gambia | 1 000 | 1 449 | 144.9 | 1.9 | 73.9 | 344 | 8.3 |
| Ghana | 22 754 | 21 053 | 92.5 | 1.8 | 54.2 | 285 | 5.2 |
| Guinea | 24 572 | 8 073 | 32.9 | 2.1 | 64.3 | 433 | 2.6 |
| Guinea-Bissau | 2 812 | 1 533 | 54.5 | 2.9 | 65.2 | 137 | 4.3 |
| Liberia | 9 632 | 3 449 | 35.8 | 2.2 | 52.7 | 120 | 2.0 |
| Mali | 122 019 | 11 937 | 9.8 | 2.4 | 67.0 | 260 | 2.2 |
| Niger | 126 670 | 12 095 | 9.6 | 2.8 | 77.3 | 174 | 0.9 |
| Nigeria | 91 077 | 139 824 | 153.5 | 2.4 | 52.5 | 361 | 3.6 |
| Senegal | 19 253 | 10 455 | 54.3 | 2.1 | 49.7 | 504 | 6.0 |
| Sierra Leone | 7 162 | 5 436 | 75.9 | 1.9 | 60.5 | 206 | 7.4 |
| Togo | 5 439 | 4 966 | 91.3 | 2.1 | 64.3 | 294 | 3.0 |
| **Total West Africa** | **503 015** | **257 170** | | | | | |
| **Total Africa** | **2 962 656** | **868 182** | | | | | |
| | | | | | | | |
| Afghanistan | 65 209 | 17 685 | 45.3 | – | 76.2 | – | 7.5 |
| Armenia | 2 820 | 3 050 | 108.1 | −0.2 | 35.7 | 975 | 10.1 |
| Azerbaijan | 8 260 | 8 280 | 100.2 | 0.6 | 50.0 | 957 | 11.2 |
| Georgia | 6 949 | 4 521 | 65.1 | −1.0 | 48.3 | 897 | 8.5 |
| Kazakhstan | 269 970 | 14 958 | 5.5 | 0.5 | 44.1 | 1 822 | 9.4 |
| Kyrgyzstan | 19 180 | 5 099 | 26.6 | 0.9 | 66.1 | 324 | 7.1 |
| Tajikistan | 14 060 | 6 430 | 45.7 | 0.7 | 75.5 | 226 | 10.6 |
| Turkmenistan | 46 993 | 4 931 | 10.5 | 1.4 | 54.4 | 1 142 | 17.0 |
| Uzbekistan | 41 424 | 25 930 | 62.6 | 1.3 | 63.5 | 645 | 7.7 |
| **Total Central Asia** | **474 865** | **90 884** | | | | | |

**NOTE:** The regional breakdown reflects geographic rather than economic or political groupings.

TABLE 1 (CONT.)
**Basic data on countries and areas**

| Country/area | Land area[a] | Population 2004 | | | | GDP 2004 | |
|---|---|---|---|---|---|---|---|
| | | Total | Density | Annual growth rate | Rural | Per capita | Annual growth rate |
| | *(1 000 ha)* | *(1 000)* | *(Population/ km²)* | *(%)* | *(% of total)* | *(US$)* | *(%)* |
| China | 932 742 | 1 326 544 | 142.2 | 0.6 | 60.4 | 1 162 | 9.5 |
| Democratic People's Republic of Korea | 12 041 | 22 745 | 188.9 | 0.6 | 38.6 | – | – |
| Japan | 36 450 | 127 764 | 350.5 | 0.2 | 34.4 | 39 195 | 2.7 |
| Mongolia | 156 650 | 2 515 | 1.6 | 1.4 | 43.1 | 462 | 10.6 |
| Republic of Korea | 9 873 | 48 142 | 487.6 | 0.5 | 19.5 | 12 743 | 4.6 |
| **Total East Asia** | **1 147 756** | **1 527 710** | | | | | |
| | | | | | | | |
| Bangladesh | 13 017 | 140 494 | 1 079.3 | 1.7 | 75.4 | 396 | 5.5 |
| Bhutan | 4 700 | 896 | 19.1 | 2.5 | 91.2 | 695 | 4.9 |
| India | 297 319 | 1 079 721 | 363.2 | 1.4 | 71.5 | 538 | 6.9 |
| Maldives | 30 | 300 | 998.4 | 2.2 | 70.7 | 2 693 | 8.8 |
| Nepal | 14 300 | 25 190 | 176.2 | 2.1 | 84.6 | 245 | 3.7 |
| Pakistan | 77 088 | 152 061 | 197.3 | 2.4 | 65.5 | 566 | 6.4 |
| Sri Lanka | 6 463 | 19 444 | 300.9 | 1.1 | 79.0 | 965 | 6.0 |
| **Total South Asia** | **412 917** | **1 418 106** | | | | | |
| | | | | | | | |
| Brunei Darussalam | 527 | 361 | 68.6 | 1.4 | 23.2 | – | – |
| Cambodia | 17 652 | 13 630 | 77.2 | 1.7 | 80.8 | 328 | 6.0 |
| Indonesia | 181 157 | 217 588 | 120.1 | 1.4 | 53.3 | 906 | 5.1 |
| Lao People's Democratic Republic | 23 080 | 5 792 | 25.1 | 2.3 | 78.8 | 372 | 6.0 |
| Malaysia | 32 855 | 25 209 | 76.7 | 1.7 | 35.6 | 4 221 | 7.1 |
| Myanmar | 65 755 | 49 910 | 75.9 | 1.1 | 70.0 | – | – |
| Philippines | 29 817 | 82 987 | 278.3 | 1.8 | 38.2 | 1 079 | 6.2 |
| Singapore | 67 | 4 335 | 6 470.2 | 2.0 | 0.0 | 23 636 | 8.4 |
| Thailand | 51 089 | 62 387 | 122.1 | 0.6 | 67.8 | 2 399 | 6.1 |
| Timor-Leste | 1 487 | 925 | 62.2 | 5.3 | 92.3 | 355 | 1.8 |
| Viet Nam | 32 549 | 82 162 | 252.4 | 1.0 | 73.8 | 500 | 7.5 |
| **Total Southeast Asia** | **436 035** | **545 286** | | | | | |
| | | | | | | | |
| Bahrain | 71 | 725 | 1 021.7 | 1.9 | 9.9 | – | – |
| Cyprus | 924 | 776 | 83.9 | 0.7 | 30.7 | 13 245 | 3.7 |
| Iran (Islamic Republic of) | 163 620 | 66 928 | 40.9 | 1.3 | 32.7 | 1 812 | 6.5 |
| Iraq | 43 737 | 24 700 | 57.8 | 2.2 | 33.0 | – | – |
| Israel | 2 171 | 6 798 | 313.1 | 1.6 | 8.3 | 17 752 | 4.3 |
| Jordan | 8 893 | 5 440 | 61.2 | 2.5 | 20.8 | 1 908 | 7.5 |
| Kuwait | 1 782 | 2 460 | 138.0 | 2.6 | 3.7 | – | – |
| Lebanon | 1 023 | 4 554 | 445.2 | 1.2 | 12.3 | 4 358 | 6.3 |

**NOTE:** The regional breakdown reflects geographic rather than economic or political groupings.

| Country/area | Land area[a] | Population 2004 | | | | GDP 2004 | |
|---|---|---|---|---|---|---|---|
| | | Total | Density | Annual growth rate | Rural | Per capita | Annual growth rate |
| | (1 000 ha) | (1 000) | (Population/ km²) | (%) | (% of total) | (US$) | (%) |
| Occupied Palestinian Territory | 602 | 3 508 | 564.0 | 4.1 | – | – | – |
| Oman | 30 950 | 2 659 | 8.6 | 2.3 | 21.9 | – | – |
| Qatar | 1 100 | 637 | 57.9 | 2.1 | 7.8 | – | – |
| Saudi Arabia | 214 969 | 23 215 | 10.8 | 3.0 | 12.0 | 9 259 | 5.2 |
| Syrian Arab Republic | 18 378 | 17 783 | 96.8 | 2.3 | 49.8 | 1 150 | 3.6 |
| Turkey | 76 963 | 71 727 | 93.2 | 1.4 | 33.2 | 3 197 | 8.9 |
| United Arab Emirates | 8 360 | 4 284 | 51.2 | – | 14.7 | – | – |
| Yemen | 52 797 | 19 763 | 37.4 | 3.0 | 74.0 | 550 | 2.7 |
| **Total Western Asia** | **626 340** | **255 957** | | | | | |
| **Total Asia** | **3 097 913** | **3 837 943** | | | | | |
| | | | | | | | |
| Albania | 2 740 | 3 188 | 116.4 | 0.6 | 55.6 | 1 470 | 6.2 |
| Andorra | 48 | 66 | 141.0 | – | – | – | – |
| Austria | 8 273 | 8 115 | 98.1 | 0.3 | 34.2 | 24 674 | 2.2 |
| Belarus | 20 748 | 9 832 | 47.4 | –0.5 | 28.7 | 1 516 | 11.0 |
| Belgium | 3 028 | 10 405 | 344.2 | 0.3 | 2.8 | 23 134 | 2.9 |
| Bosnia and Herzegovina | 5 120 | 3 836 | 74.9 | 0.0 | 55.2 | 1 384 | 4.7 |
| Bulgaria | 11 063 | 7 780 | 70.3 | –0.6 | 29.9 | 1 951 | 5.6 |
| Channel Islands | 19 | 149 | 745.0 | 0.0 | 69.5 | – | – |
| Croatia | 5 592 | 4 508 | 80.6 | 0.1 | 40.6 | 4 857 | 3.7 |
| Czech Republic | 7 728 | 10 183 | 131.8 | –0.2 | 25.6 | 6 148 | 4.0 |
| Denmark | 4 243 | 5 397 | 127.2 | 0.2 | 14.5 | 30 930 | 2.4 |
| Estonia | 4 239 | 1 345 | 31.7 | –0.6 | 30.5 | 5 170 | 6.2 |
| Faeroe Islands | 140 | 48 | 34.3 | – | – | – | – |
| Finland | 30 459 | 5 215 | 17.1 | 0.1 | 39.1 | 25 107 | 3.7 |
| France | 55 010 | 59 991 | 109.1 | 0.4 | 23.5 | 23 157 | 2.3 |
| Germany | 34 895 | 82 631 | 236.8 | 0.1 | 11.7 | 23 209 | 1.6 |
| Gibraltar | 1 | 28 | 2 788.4 | – | – | – | – |
| Greece | 12 890 | 11 075 | 85.9 | 0.4 | 38.9 | 11 885 | 4.2 |
| Holy See | | 1 | | – | – | – | – |
| Hungary | 9 210 | 10 072 | 109.4 | –0.6 | 34.5 | 5 339 | 4.0 |
| Iceland | 10 025 | 290 | 2.9 | 0.4 | 7.1 | 32 449 | 5.2 |
| Ireland | 6 889 | 4 019 | 58.3 | 0.6 | 39.9 | 29 118 | 4.9 |
| Isle of Man | 57 | 77 | 134.6 | – | – | – | – |
| Italy | 29 411 | 57 573 | 195.8 | –0.1 | 32.5 | 19 344 | 1.2 |
| Latvia | 6 205 | 2 303 | 37.1 | –0.8 | 33.9 | 4 502 | 8.5 |
| Liechtenstein | 16 | 34 | 212.5 | – | – | – | – |
| Lithuania | 6 268 | 3 439 | 54.9 | –0.4 | 33.3 | 4 398 | 6.7 |
| Luxembourg | 259 | 450 | 174.0 | 0.5 | 7.9 | 47 926 | 4.5 |
| Malta | 32 | 401 | 1 253.1 | 0.5 | 8.1 | 9 508 | 1.4 |

**NOTE:** The regional breakdown reflects geographic rather than economic or political groupings.

TABLE 1 (CONT.)

## Basic data on countries and areas

| Country/area | Land area[a] | Population 2004 | | | | GDP 2004 | |
|---|---|---|---|---|---|---|---|
| | | Total | Density | Annual growth rate | Rural | Per capita | Annual growth rate |
| | (1 000 ha) | (1 000) | (Population/ km$^2$) | (%) | (% of total) | (US$) | (%) |
| Monaco | 2 | 33 | 16 923.1 | – | – | – | – |
| Netherlands | 3 388 | 16 250 | 479.6 | 0.2 | 33.7 | 23 255 | 1.4 |
| Norway | 30 625 | 4 582 | 15.0 | 0.4 | 20.5 | 39 198 | 2.9 |
| Poland | 30 629 | 38 160 | 124.6 | −0.1 | 38.1 | 4 885 | 5.3 |
| Portugal | 9 150 | 10 436 | 114.1 | 0.7 | 44.9 | 10 395 | 1.0 |
| Republic of Moldova | 3 288 | 4 218 | 128.3 | −0.5 | 53.8 | 398 | 7.3 |
| Romania | 22 987 | 21 858 | 95.1 | −0.3 | 45.3 | 2 115 | 8.3 |
| Russian Federation | 1 688 850 | 142 814 | 8.5 | −0.4 | 26.7 | 2 302 | 7.2 |
| San Marino | 6 | 28 | 462.8 | – | – | – | – |
| Serbia and Montenegro | 10 200 | 8 152 | 79.9 | −0.7 | 47.8 | 1 272 | 7.2 |
| Slovakia | 4 808 | 5 390 | 110.5 | 0.0 | 42.3 | 4 488 | 5.5 |
| Slovenia | 2 012 | 1 995 | 99.2 | 0.0 | 49.2 | 10 871 | 4.6 |
| Spain | 49 944 | 41 286 | 82.7 | 0.5 | 23.4 | 15 079 | 3.1 |
| Sweden | 41 162 | 8 985 | 21.8 | 0.3 | 16.6 | 28 912 | 3.6 |
| Switzerland | 3 955 | 7 382 | 186.7 | 0.4 | 32.5 | 34 190 | 1.7 |
| The former Yugoslav Republic of Macedonia | 2 543 | 2 062 | 81.1 | 0.6 | 40.4 | 1 772 | 2.5 |
| Ukraine | 57 935 | 48 008 | 82.9 | −0.7 | 32.7 | 917 | 12.1 |
| United Kingdom | 24 088 | 59 405 | 246.6 | 0.1 | 10.8 | 26 506 | 3.1 |
| **Total Europe** | **2 260 180** | **723 495** | | | | | |
| | | | | | | | |
| Anguilla | 8 | 13 | 172.1 | – | – | – | – |
| Antigua and Barbuda | 44 | 80 | 181.8 | 2.7 | 61.9 | 9 608 | 4.1 |
| Aruba | 19 | 99 | 521.1 | – | – | – | – |
| Bahamas | 1 001 | 320 | 32.0 | 0.8 | 10.3 | – | – |
| Barbados | 43 | 272 | 632.1 | 0.4 | 47.7 | – | – |
| Bermuda | 5 | 64 | 1 280.0 | 0.3 | 0.0 | – | – |
| British Virgin Islands | 15 | 23 | 151.0 | – | – | – | – |
| Cayman Islands | 26 | 44 | 169.2 | – | 57.3 | – | – |
| Cuba | 10 982 | 11 365 | 103.5 | 0.3 | 24.2 | – | – |
| Dominica | 75 | 71 | 95.3 | 0.4 | 27.6 | 3 534 | 2.0 |
| Dominican Republic | 4 838 | 8 861 | 183.2 | 1.4 | 40.3 | 2 450 | 2.0 |
| Grenada | 34 | 106 | 310.9 | 1.1 | 58.6 | 3 798 | −2.8 |
| Guadeloupe | 169 | 449 | 265.5 | – | – | – | – |
| Haiti | 2 756 | 8 592 | 311.8 | 1.8 | 61.9 | 437 | −3.8 |
| Jamaica | 1 083 | 2 665 | 246.1 | 0.8 | 47.8 | 2 975 | 2.0 |
| Martinique | 106 | 433 | 408.4 | – | – | – | – |
| Montserrat | 10 | 9 | 93.4 | – | – | – | – |

**NOTE:** The regional breakdown reflects geographic rather than economic or political groupings.

| Country/area | Land area[a] | Population 2004 | | | | GDP 2004 | |
|---|---|---|---|---|---|---|---|
| | | Total | Density | Annual growth rate | Rural | Per capita | Annual growth rate |
| | *(1 000 ha)* | *(1 000)* | *(Population/ km²)* | *(%)* | *(% of total)* | *(US$)* | *(%)* |
| Netherlands Antilles | 80 | 222 | 277.5 | 0.8 | 30.1 | – | – |
| Puerto Rico | 887 | 3 929 | 442.9 | 0.8 | 3.1 | – | – |
| Saint Kitts and Nevis | 36 | 47 | 130.5 | 0.6 | 68.0 | 7 427 | 4.0 |
| Saint Lucia | 61 | 164 | 268.3 | 1.9 | 69.1 | 4 276 | 3.5 |
| Saint Vincent and the Grenadines | 39 | 108 | 277.7 | –0.8 | 40.7 | 3 382 | 4.0 |
| Trinidad and Tobago | 513 | 1 323 | 258.0 | 0.8 | 24.2 | 7 921 | 6.2 |
| Turks and Caicos Islands | 43 | 21 | 47.8 | – | – | – | – |
| United States Virgin Islands | 34 | 113 | 332.8 | 1.4 | 6.2 | – | – |
| **Total Caribbean** | **22 907** | **39 393** | | | | | |
| | | | | | | | |
| Belize | 2 280 | 283 | 12.4 | 3.2 | 51.5 | 3 669 | 4.2 |
| Costa Rica | 5 106 | 4 061 | 79.5 | 1.4 | 38.8 | 4 534 | 4.2 |
| El Salvador | 2 072 | 6 658 | 321.3 | 1.9 | 40.2 | 2 124 | 1.7 |
| Guatemala | 10 843 | 12 628 | 116.5 | 2.6 | 53.3 | 1 676 | 2.7 |
| Honduras | 11 189 | 7 141 | 63.8 | 2.5 | 54.0 | 952 | 4.6 |
| Nicaragua | 12 140 | 5 604 | 46.2 | 2.2 | 42.3 | 778 | 3.7 |
| Panama | 7 443 | 3 028 | 40.7 | 1.5 | 42.5 | 4 373 | 6.2 |
| **Total Central America** | **51 073** | **39 403** | | | | | |
| | | | | | | | |
| Argentina | 273 669 | 38 226 | 14.0 | 0.8 | 9.7 | 7 511 | 9.0 |
| Bolivia | 108 438 | 8 986 | 8.3 | 1.9 | 36.1 | 1 036 | 3.6 |
| Brazil | 845 942 | 178 718 | 21.1 | 1.2 | 16.4 | 3 675 | 5.2 |
| Chile | 74 880 | 15 956 | 21.3 | 1.2 | 12.7 | 5 448 | 6.1 |
| Colombia | 103 870 | 45 300 | 43.6 | 1.6 | 23.1 | 2 069 | 4.0 |
| Ecuador | 27 684 | 13 213 | 47.7 | 1.6 | 37.7 | 1 435 | 6.6 |
| Falkland Islands | 1 217 | 3 | 0.2 | – | – | – | – |
| French Guiana | 8 815 | 196 | 2.2 | – | – | – | – |
| Guyana | 19 685 | 772 | 3.9 | 0.4 | 62.0 | 962 | 1.6 |
| Paraguay | 39 730 | 5 782 | 14.6 | 2.4 | 42.1 | 1 413 | 2.9 |
| Peru | 128 000 | 27 547 | 21.5 | 1.5 | 25.8 | 2 207 | 5.1 |
| South Georgia and the South Sandwich Islands | 409 | 0 | | – | – | – | – |
| Suriname | 15 600 | 443 | 2.8 | 1.1 | 23.4 | 2 388 | 4.6 |
| Uruguay | 17 502 | 3 399 | 19.4 | 0.6 | 7.3 | 5 826 | 12.3 |
| Venezuela (Bolivarian Republic of) | 88 205 | 26 127 | 29.6 | 1.8 | 12.1 | 4 575 | 17.3 |
| **Total South America** | **1 753 646** | **364 668** | | | | | |
| **Total Latin America and the Caribbean** | **1 827 626** | **443 464** | | | | | |

**NOTE:** The regional breakdown reflects geographic rather than economic or political groupings.

TABLE 1 (CONT.)

## Basic data on countries and areas

| Country/area | Land area[a] | Population 2004 | | | | GDP 2004 | |
|---|---|---|---|---|---|---|---|
| | | Total | Density | Annual growth rate | Rural | Per capita | Annual growth rate |
| | (1 000 ha) | (1 000) | (Population/ km²) | (%) | (% of total) | (US$) | (%) |
| Canada | 922 097 | 31 902 | 3.5 | 0.9 | 19.2 | 24 712 | 2.9 |
| Greenland | 41 045 | 57 | 0.1 | –0.4 | 17.3 | – | – |
| Mexico | 190 869 | 103 795 | 54.4 | 1.5 | 24.2 | 5 968 | 4.4 |
| Saint Pierre and Miquelon | 23 | 7 | 30.5 | – | – | – | – |
| United States of America | 915 896 | 293 507 | 32.1 | 0.9 | 19.6 | 36 790 | 4.4 |
| **Total North America** | **2 069 930** | **429 268** | | | | | |
| | | | | | | | |
| American Samoa | 20 | 57 | 285.0 | – | – | – | – |
| Australia | 768 230 | 20 120 | 2.6 | 1.2 | 7.7 | 22 074 | 3.0 |
| Cook Islands | 23 | 21 | 93.0 | – | – | – | – |
| Fiji | 1 827 | 848 | 46.4 | 1.5 | 47.5 | 2 232 | 3.8 |
| French Polynesia | 366 | 246 | 67.2 | 1.2 | 47.9 | – | – |
| Guam | 55 | 164 | 298.0 | 1.4 | 6.2 | – | – |
| Kiribati | 73 | 98 | 134.0 | 1.5 | 51.3 | 532 | 1.8 |
| Marshall Islands | 18 | 60 | 330.9 | | 33.6 | 1 738 | 1.5 |
| Micronesia (Federated States of) | 70 | 127 | 180.6 | 1.8 | 70.3 | 1 745 | –3.8 |
| Nauru | 2 | 13 | 652.4 | – | – | – | – |
| New Caledonia | 1 828 | 229 | 12.5 | 1.9 | 38.6 | – | – |
| New Zealand | 26 799 | 4 061 | 15.2 | 1.3 | 14.1 | 14 984 | 4.4 |
| Niue | 26 | 2 | 8.3 | – | – | – | – |
| Northern Mariana Islands | 46 | 77 | 161.4 | – | – | – | – |
| Palau | 46 | 20 | 43.5 | | 31.6 | 6 360 | 2.0 |
| Papua New Guinea | 45 286 | 5 625 | 12.4 | 2.2 | 86.8 | 622 | 2.8 |
| Pitcairn | 4 | 0 | | – | – | – | – |
| Samoa | 283 | 179 | 63.3 | 0.6 | 77.6 | 1 417 | 3.2 |
| Solomon Islands | 2 799 | 471 | 16.8 | 3.1 | 83.2 | 621 | 3.8 |
| Tokelau | 1 | 1 | 139.2 | – | – | – | – |
| Tonga | 72 | 102 | 141.4 | 0.3 | 66.2 | 1 638 | 1.6 |
| Tuvalu | 3 | 12 | | – | – | – | – |
| Vanuatu | 1 219 | 215 | 17.6 | 2.3 | 76.7 | 1 110 | 3.0 |
| Wallis and Futuna Islands | 20 | 16 | 80.1 | – | – | – | – |
| **Total Oceania** | **849 116** | **32 764** | | | | | |
| | | | | | | | |
| **Total World** | **13 067 421** | **6 335 116** | | | | | |

[a] "Land area" refers to the total area of a country, excluding areas under inland water bodies. The world total corresponds to the sum of the reporting units; about 35 million hectares of land in Antarctica, some Arctic and Antarctic islands and some other minor islands are not included.

**SOURCE:** FAO, 2006a. Economic and demographic figures are from World Bank, 2006a.

**NOTE:** The regional breakdown reflects geographic rather than economic or political groupings.

TABLE 2
## Forest area and area change

| Country/area | Forest area, 2005 | | | | Annual change rate | | | |
|---|---|---|---|---|---|---|---|---|
| | Total forest[a] | % of land area | Area per capita | Forest plantations | 1990–2000 | | 2000–2005 | |
| | (1 000 ha) | (%) | (ha) | (1 000 ha) | (1 000 ha) | (%) | (1 000 ha) | (%) |
| Burundi | 152 | 5.9 | 0.0 | 86 | –9 | –3.7 | –9 | –5.2 |
| Cameroon | 21 245 | 45.6 | 1.3 | – | –220 | –0.9 | –220 | –1 |
| Central African Republic | 22 755 | 36.5 | 5.8 | – | –30 | –0.1 | –30 | –0.1 |
| Chad | 11 921 | 9.5 | 1.4 | – | –79 | –0.6 | –79 | –0.7 |
| Congo | 22 471 | 65.8 | 5.8 | – | –17 | –0.1 | –17 | –0.1 |
| Democratic Republic of the Congo | 133 610 | 58.9 | 2.4 | – | –532 | –0.4 | –319 | –0.2 |
| Equatorial Guinea | 1 632 | 58.2 | 3.2 | – | –15 | –0.8 | –15 | –0.9 |
| Gabon | 21 775 | 84.5 | 15.8 | – | –10 | n.s. | –10 | n.s. |
| Rwanda | 480 | 19.5 | 0.1 | 419 | 3 | 0.8 | 27 | 6.9 |
| Saint Helena | 2 | 6.5 | 0.3 | – | 0 | 0 | 0 | 0 |
| Sao Tome and Principe | 27 | 28.4 | 0.2 | – | 0 | 0 | 0 | 0 |
| **Total Central Africa** | **236 070** | **44.6** | **2.2** | | **–910** | **–0.37** | **–673** | **–0.28** |
| | | | | | | | | |
| British Indian Ocean Territory | 3 | 32.5 | 3.0 | – | 0 | 0 | 0 | 0 |
| Comoros | 5 | 2.9 | 0.0 | – | n.s. | –4 | –1 | –7.4 |
| Djibouti | 6 | 0.2 | 0.0 | – | 0 | 0 | 0 | 0 |
| Eritrea | 1 554 | 15.4 | 0.3 | 28 | –4 | –0.3 | –4 | –0.3 |
| Ethiopia | 13 000 | 11.9 | 0.2 | – | –141 | –1 | –141 | –1.1 |
| Kenya | 3 522 | 6.2 | 0.1 | – | –13 | –0.3 | –12 | –0.3 |
| Madagascar | 12 838 | 22.1 | 0.7 | 293 | –67 | –0.5 | –37 | –0.3 |
| Mauritius | 37 | 18.2 | 0.0 | 15 | n.s. | –0.3 | n.s. | –0.5 |
| Mayotte | 5 | 14.7 | 0.0 | – | n.s. | –0.4 | n.s. | –0.4 |
| Réunion | 84 | 33.6 | 0.1 | 4 | n.s. | –0.1 | –1 | –0.7 |
| Seychelles | 40 | 88.9 | 0.5 | – | 0 | 0 | 0 | 0 |
| Somalia | 7 131 | 11.4 | 0.7 | – | –77 | –1 | –77 | –1 |
| Uganda | 3 627 | 18.4 | 0.1 | – | –86 | –1.9 | –86 | –2.2 |
| United Republic of Tanzania | 35 257 | 39.9 | 1.0 | – | –412 | –1 | –412 | –1.1 |
| **Total East Africa** | **77 109** | **18.9** | **0.4** | | **–801** | **–0.94** | **–771** | **–0.97** |
| | | | | | | | | |
| Algeria | 2 277 | 1 | 0.1 | 754 | 35 | 1.8 | 27 | 1.2 |
| Egypt | 67 | 0.1 | 0.0 | 67 | 2 | 3 | 2 | 2.6 |
| Libyan Arab Jamahiriya | 217 | 0.1 | 0.0 | – | 0 | 0 | 0 | 0 |
| Mauritania | 267 | 0.3 | 0.1 | – | –10 | –2.7 | –10 | –3.4 |
| Morocco | 4 364 | 9.8 | 0.1 | – | 4 | 0.1 | 7 | 0.2 |
| Sudan | 67 546 | 28.4 | 2.0 | 5 403 | –589 | –0.8 | –589 | –0.8 |
| Tunisia | 1 056 | 6.8 | 0.1 | 498 | 32 | 4.1 | 19 | 1.9 |
| Western Sahara | 1 011 | 3.8 | 3.7 | – | 0 | 0 | 0 | 0 |
| **Total Northern Africa** | **76 805** | **8.2** | **0.4** | | **–526** | **–0.64** | **–544** | **–0.69** |

n.s. = not significant, indicating a very small value
– = not available
0 = true zero
**NOTE:** The regional breakdown reflects geographic rather than economic or political groupings.

TABLE 2 (CONT.)
## Forest area and area change

| Country/area | Forest area, 2005 | | | | Annual change rate | | | |
|---|---|---|---|---|---|---|---|---|
| | Total forest[a] | % of land area | Area per capita | Forest plantations | 1990–2000 | | 2000–2005 | |
| | (1 000 ha) | (%) | (ha) | (1 000 ha) | (1 000 ha) | (%) | (1 000 ha) | (%) |
| Angola | 59 104 | 47.4 | 4.2 | 131 | –125 | –0.2 | –125 | –0.2 |
| Botswana | 11 943 | 21.1 | 6.9 | – | –118 | –0.9 | –118 | –1 |
| Lesotho | 8 | 0.3 | 0.0 | – | n.s. | 3.4 | n.s. | 2.7 |
| Malawi | 3 402 | 36.2 | 0.3 | – | –33 | –0.9 | –33 | –0.9 |
| Mozambique | 19 262 | 24.6 | 1.0 | – | –50 | –0.3 | –50 | –0.3 |
| Namibia | 7 661 | 9.3 | 3.8 | – | –73 | –0.9 | –74 | –0.9 |
| South Africa | 9 203 | 7.6 | 0.2 | – | 0 | 0 | 0 | 0 |
| Swaziland | 541 | 31.5 | 0.5 | – | 5 | 0.9 | 5 | 0.9 |
| Zambia | 42 452 | 57.1 | 4.0 | – | –445 | –0.9 | –445 | –1 |
| Zimbabwe | 17 540 | 45.3 | 1.3 | – | –313 | –1.5 | –313 | –1.7 |
| **Total Southern Africa** | **171 116** | **29** | **1.4** | | **–1 152** | **–0.63** | **–1 154** | **–0.66** |
| | | | | | | | | |
| Benin | 2 351 | 21.3 | 0.3 | – | –65 | –2.1 | –65 | –2.5 |
| Burkina Faso | 6 794 | 29 | 0.5 | 76 | –24 | –0.3 | –24 | –0.3 |
| Cape Verde | 84 | 20.7 | 0.2 | 84 | 2 | 3.6 | n.s. | 0.4 |
| Côte d'Ivoire | 10 405 | 32.7 | 0.6 | 337 | 11 | 0.1 | 15 | 0.1 |
| Gambia | 471 | 41.7 | 0.3 | – | 2 | 0.4 | 2 | 0.4 |
| Ghana | 5 517 | 24.2 | 0.3 | – | –135 | –2 | –115 | –2 |
| Guinea | 6 724 | 27.4 | 0.8 | 33 | –50 | –0.7 | –36 | –0.5 |
| Guinea-Bissau | 2 072 | 73.7 | 1.4 | – | –10 | –0.4 | –10 | –0.5 |
| Liberia | 3 154 | 32.7 | 0.9 | – | –60 | –1.6 | –60 | –1.8 |
| Mali | 12 572 | 10.3 | 1.1 | – | –100 | –0.7 | –100 | –0.8 |
| Niger | 1 266 | 1 | 0.1 | – | –62 | –3.7 | –12 | –1 |
| Nigeria | 11 089 | 12.2 | 0.1 | 349 | –410 | –2.7 | –410 | –3.3 |
| Senegal | 8 673 | 45 | 0.8 | 365 | –45 | –0.5 | –45 | –0.5 |
| Sierra Leone | 2 754 | 38.5 | 0.5 | – | –19 | –0.7 | –19 | –0.7 |
| Togo | 386 | 7.1 | 0.1 | 38 | –20 | –3.4 | –20 | –4.5 |
| **Total West Africa** | **74 312** | **14.9** | **0.3** | | **–985** | **–1.17** | **–899** | **–1.17** |
| **Total Africa** | **635 412** | **21.4** | **0.7** | | **–4 375** | **–0.64** | **–4 040** | **–0.62** |
| | | | | | | | | |
| Afghanistan | 867 | 1.3 | 0.0 | – | –29 | –2.5 | –30 | –3.1 |
| Armenia | 283 | 10 | 0.1 | 10 | –4 | –1.3 | –4 | –1.5 |
| Azerbaijan | 936 | 11.3 | 0.1 | – | 0 | 0 | 0 | 0 |
| Georgia | 2 760 | 39.7 | 0.6 | 60 | n.s. | n.s. | n.s. | n.s. |
| Kazakhstan | 3 337 | 1.2 | 0.2 | 909 | –6 | –0.2 | –6 | –0.2 |
| Kyrgyzstan | 869 | 4.5 | 0.2 | 66 | 2 | 0.3 | 2 | 0.3 |
| Tajikistan | 410 | 2.9 | 0.1 | 66 | n.s. | n.s. | 0 | 0 |
| Turkmenistan | 4 127 | 8.8 | 0.8 | – | 0 | 0 | 0 | 0 |

**NOTE:** The regional breakdown reflects geographic rather than economic or political groupings.

| Country/area | Forest area, 2005 | | | | Annual change rate | | | |
|---|---|---|---|---|---|---|---|---|
| | Total forest[a] | % of land area | Area per capita | Forest plantations | 1990–2000 | | 2000–2005 | |
| | (1 000 ha) | (%) | (ha) | (1 000 ha) | (1 000 ha) | (%) | (1 000 ha) | (%) |
| Uzbekistan | 3 295 | 8 | 0.1 | 61 | 17 | 0.5 | 17 | 0.5 |
| **Total Central Asia** | **16 884** | **3.6** | **0.2** | | **–20** | **–0.12** | **–21** | **–0.12** |
| China | 197 290 | 21.2 | 0.1 | 31369 | 1 986 | 1.2 | 4 058 | 2.2 |
| Democratic People's Republic of Korea | 6 187 | 51.4 | 0.3 | – | –138 | –1.8 | –127 | –1.9 |
| Japan | 24 868 | 68.2 | 0.2 | – | –7 | n.s. | –2 | n.s. |
| Mongolia | 10 252 | 6.5 | 4.1 | – | –83 | –0.7 | –83 | –0.8 |
| Republic of Korea | 6 265 | 63.5 | 0.1 | – | –7 | –0.1 | –7 | –0.1 |
| **Total East Asia** | **244 862** | **21.3** | **0.2** | | **1 751** | **0.81** | **3 840** | **1.65** |
| Bangladesh | 871 | 6.7 | 0.0 | 279 | n.s. | n.s. | –2 | –0.3 |
| Bhutan | 3 195 | 68 | 3.6 | 2 | 11 | 0.3 | 11 | 0.3 |
| India | 67 701 | 22.8 | 0.1 | 3 226 | 362 | 0.6 | 29 | n.s. |
| Maldives | 1 | 3 | 0.0 | – | 0 | 0 | 0 | 0 |
| Nepal | 3 636 | 25.4 | 0.1 | 53 | –92 | –2.1 | –53 | –1.4 |
| Pakistan | 1 902 | 2.5 | 0.0 | – | –41 | –1.8 | –43 | –2.1 |
| Sri Lanka | 1 933 | 29.9 | 0.1 | 195 | –27 | –1.2 | –30 | –1.5 |
| **Total South Asia** | **79 239** | **19.2** | **0.1** | | **213** | **0.27** | **–88** | **–0.11** |
| Brunei Darussalam | 278 | 52.8 | 0.8 | – | –2 | –0.8 | –2 | –0.7 |
| Cambodia | 10 447 | 59.2 | 0.8 | – | –140 | –1.1 | –219 | –2 |
| Indonesia | 88 495 | 48.8 | 0.4 | – | –1 872 | –1.7 | –1 871 | –2 |
| Lao People's Democratic Republic | 16 142 | 69.9 | 2.8 | 224 | –78 | –0.5 | –78 | –0.5 |
| Malaysia | 20 890 | 63.6 | 0.8 | – | –78 | –0.4 | –140 | –0.7 |
| Myanmar | 32 222 | 49 | 0.6 | 849 | –466 | –1.3 | –466 | –1.4 |
| Philippines | 7 162 | 24 | 0.1 | 620 | –262 | –2.8 | –157 | –2.1 |
| Singapore | 2 | 3.4 | 0.0 | 0 | 0 | 0 | 0 | 0 |
| Thailand | 14 520 | 28.4 | 0.2 | 3 099 | –115 | –0.7 | –59 | –0.4 |
| Timor-Leste | 798 | 53.7 | 0.9 | – | –11 | –1.2 | –11 | –1.3 |
| Viet Nam | 12 931 | 39.7 | 0.2 | 2 695 | 236 | 2.3 | 241 | 2 |
| **Total Southeast Asia** | **203 887** | **46.8** | **0.4** | | **–2 790** | **–1.2** | **–2 763** | **–1.3** |
| Bahrain | n.s. | 0.6 | n.s. | – | n.s. | 5.6 | n.s. | 3.8 |
| Cyprus | 174 | 18.9 | 0.2 | 5 | 1 | 0.7 | n.s. | 0.2 |
| Iran (Islamic Republic of) | 11 075 | 6.8 | 0.2 | – | 0 | 0 | 0 | 0 |
| Iraq | 822 | 1.9 | 0.0 | 13 | 1 | 0.2 | 1 | 0.1 |
| Israel | 171 | 8.3 | 0.0 | – | 1 | 0.6 | 1 | 0.8 |
| Jordan | 83 | 0.9 | 0.0 | 40 | 0 | 0 | 0 | 0 |
| Kuwait | 6 | 0.3 | 0.0 | – | n.s. | 3.5 | n.s. | 2.7 |
| Lebanon | 136 | 13.3 | 0.0 | 8 | 1 | 0.8 | 1 | 0.8 |

**NOTE:** The regional breakdown reflects geographic rather than economic or political groupings.

TABLE 2 (CONT.)

## Forest area and area change

| Country/area | Forest area, 2005 | | | | Annual change rate | | | |
|---|---|---|---|---|---|---|---|---|
| | Total forest[a] | % of land area | Area per capita | Forest plantations | 1990–2000 | | 2000–2005 | |
| | (1 000 ha) | (%) | (ha) | (1 000 ha) | (1 000 ha) | (%) | (1 000 ha) | (%) |
| Occupied Palestinian Territory | 9 | 1.5 | 0.0 | – | 0 | 0 | 0 | 0 |
| Oman | 2 | n.s. | 0.0 | – | 0 | 0 | 0 | 0 |
| Qatar | n.s. | n.s. | n.s. | – | 0 | 0 | 0 | 0 |
| Saudi Arabia | 2 728 | 1.3 | 0.1 | – | 0 | 0 | 0 | 0 |
| Syrian Arab Republic | 461 | 2.5 | 0.0 | – | 6 | 1.5 | 6 | 1.3 |
| Turkey | 10 175 | 13.2 | 0.1 | 2 537 | 37 | 0.4 | 25 | 0.2 |
| United Arab Emirates | 312 | 3.7 | 0.1 | 312 | 6 | 2.4 | n.s. | 0.1 |
| Yemen | 549 | 1 | 0.0 | – | 0 | 0 | 0 | 0 |
| **Total Western Asia** | **26 704** | **4.3** | **0.1** | | **54** | **0.21** | **35** | **0.13** |
| **Total Asia** | **571 577** | **18.5** | **0.1** | | **–792** | **–0.14** | **1 003** | **0.18** |
| Albania | 794 | 29 | 0.2 | 89 | –2 | –0.3 | 5 | 0.6 |
| Andorra | 16 | 35.6 | 0.2 | – | 0 | 0 | 0 | 0 |
| Austria | 3 862 | 46.7 | 0.5 | – | 6 | 0.2 | 5 | 0.1 |
| Belarus | 7 894 | 38 | 0.8 | – | 47 | 0.6 | 9 | 0.1 |
| Belgium | 667 | 22 | 0.1 | 275 | –1 | –0.1 | 0 | 0 |
| Bosnia and Herzegovina | 2 185 | 43.1 | 0.6 | – | –2 | –0.1 | 0 | 0 |
| Bulgaria | 3 625 | 32.8 | 0.5 | – | 5 | 0.1 | 50 | 1.4 |
| Channel Islands | 1 | 4.1 | 0.0 | – | 0 | 0 | 0 | 0 |
| Croatia | 2 135 | 38.2 | 0.5 | 61 | 1 | 0.1 | 1 | 0.1 |
| Czech Republic | 2 648 | 34.3 | 0.3 | 0 | 1 | n.s. | 2 | 0.1 |
| Denmark | 500 | 11.8 | 0.1 | 315 | 4 | 0.9 | 3 | 0.6 |
| Estonia | 2 284 | 53.9 | 1.7 | 1 | 8 | 0.4 | 8 | 0.4 |
| Faeroe Islands | n.s. | 0.1 | n.s. | – | 0 | 0 | 0 | 0 |
| Finland | 22 500 | 73.9 | 4.3 | 0 | 28 | 0.1 | 5 | n.s. |
| France | 15 554 | 28.3 | 0.3 | – | 81 | 0.5 | 41 | 0.3 |
| Germany | 11 076 | 31.7 | 0.1 | 0 | 34 | 0.3 | 0 | 0 |
| Gibraltar | 0 | 0 | 0.0 | – | 0 | 0 | 0 | 0 |
| Greece | 3 752 | 29.1 | 0.3 | 134 | 30 | 0.9 | 30 | 0.8 |
| Holy See | 0 | 0 | 0.0 | – | 0 | 0 | 0 | 0 |
| Hungary | 1 976 | 21.5 | 0.2 | 545 | 11 | 0.6 | 14 | 0.7 |
| Iceland | 46 | 0.5 | 0.2 | 29 | 1 | 4.3 | 2 | 3.9 |
| Ireland | 669 | 9.7 | 0.2 | 579 | 17 | 3.3 | 12 | 1.9 |
| Isle of Man | 3 | 6.1 | 0.0 | – | 0 | 0 | 0 | 0 |
| Italy | 9 979 | 33.9 | 0.2 | – | 106 | 1.2 | 106 | 1.1 |
| Latvia | 2 941 | 47.4 | 1.3 | 1 | 11 | 0.4 | 11 | 0.4 |
| Liechtenstein | 7 | 43.1 | 0.2 | – | n.s. | 0.6 | 0 | 0 |

**NOTE:** The regional breakdown reflects geographic rather than economic or political groupings.

| Country/area | Forest area, 2005 | | | | Annual change rate | | | |
|---|---|---|---|---|---|---|---|---|
| | Total forest[a] | % of land area | Area per capita | Forest plantations | 1990–2000 | | 2000–2005 | |
| | (1 000 ha) | (%) | (ha) | (1 000 ha) | (1 000 ha) | (%) | (1 000 ha) | (%) |
| Lithuania | 2 099 | 33.5 | 0.6 | 141 | 8 | 0.4 | 16 | 0.8 |
| Luxembourg | 87 | 33.5 | 0.2 | – | n.s. | 0.1 | 0 | 0 |
| Malta | n.s. | 1.1 | n.s. | – | 0 | 0 | 0 | 0 |
| Monaco | 0 | 0 | 0.0 | – | 0 | 0 | 0 | 0 |
| Netherlands | 365 | 10.8 | 0.0 | 4 | 2 | 0.4 | 1 | 0.3 |
| Norway | 9 387 | 30.7 | 2.0 | – | 17 | 0.2 | 17 | 0.2 |
| Poland | 9 192 | 30 | 0.2 | – | 18 | 0.2 | 27 | 0.3 |
| Portugal | 3 783 | 41.3 | 0.4 | 1 234 | 48 | 1.5 | 40 | 1.1 |
| Republic of Moldova | 329 | 10 | 0.1 | 1 | 1 | 0.2 | 1 | 0.2 |
| Romania | 6 370 | 27.7 | 0.3 | 149 | 0 | n.s. | 1 | n.s. |
| Russian Federation | 808 790 | 47.9 | 5.7 | 16 963 | 32 | n.s. | –96 | n.s. |
| San Marino | n.s. | 1.6 | n.s. | – | 0 | 0 | 0 | 0 |
| Serbia and Montenegro | 2 694 | 26.4 | 0.3 | 39 | 9 | 0.3 | 9 | 0.3 |
| Slovakia | 1 929 | 40.1 | 0.4 | 19 | n.s. | n.s. | 2 | 0.1 |
| Slovenia | 1 264 | 62.8 | 0.6 | 0 | 5 | 0.4 | 5 | 0.4 |
| Spain | 17 915 | 35.9 | 0.4 | 1 471 | 296 | 2 | 296 | 1.7 |
| Sweden | 27 528 | 66.9 | 3.1 | 667 | 11 | n.s. | 11 | n.s. |
| Switzerland | 1 221 | 30.9 | 0.2 | 4 | 4 | 0.4 | 4 | 0.4 |
| The former Yugoslav Republic of Macedonia | 906 | 35.8 | 0.4 | – | 0 | 0 | 0 | 0 |
| Ukraine | 9 575 | 16.5 | 0.2 | 388 | 24 | 0.3 | 13 | 0.1 |
| United Kingdom | 2 845 | 11.8 | 0.0 | 1 924 | 18 | 0.7 | 10 | 0.4 |
| **Total Europe** | **1 001 394** | **44.3** | **1.4** | | **877** | **0.09** | **661** | **0.07** |
| | | | | | | | | |
| Anguilla | 6 | 71.4 | 0.5 | – | 0 | 0 | 0 | 0 |
| Antigua and Barbuda | 9 | 21.4 | 0.1 | – | 0 | 0 | 0 | 0 |
| Aruba | n.s. | 2.2 | n.s. | – | 0 | 0 | 0 | 0 |
| Bahamas | 515 | 51.5 | 1.6 | 0 | 0 | 0 | 0 | 0 |
| Barbados | 2 | 4 | 0.0 | – | 0 | 0 | 0 | 0 |
| Bermuda | 1 | 20 | 0.0 | – | 0 | 0 | 0 | 0 |
| British Virgin Islands | 4 | 24.4 | 0.2 | – | n.s. | –0.1 | n.s. | –0.1 |
| Cayman Islands | 12 | 48.4 | 0.3 | – | 0 | 0 | 0 | 0 |
| Cuba | 2 713 | 24.7 | 0.2 | 394 | 38 | 1.7 | 56 | 2.2 |
| Dominica | 46 | 61.3 | 0.6 | – | n.s. | –0.5 | n.s. | –0.6 |
| Dominican Republic | 1 376 | 28.4 | 0.2 | – | 0 | 0 | 0 | 0 |
| Grenada | 4 | 12.2 | 0.0 | – | n.s. | n.s. | 0 | 0 |
| Guadeloupe | 80 | 47.2 | 0.2 | 1 | n.s. | –0.3 | n.s. | –0.3 |
| Haiti | 105 | 3.8 | 0.0 | – | –1 | –0.6 | –1 | –0.7 |
| Jamaica | 339 | 31.3 | 0.1 | 14 | n.s. | –0.1 | n.s. | –0.1 |
| Martinique | 46 | 43.9 | 0.1 | – | 0 | 0 | 0 | 0 |

**NOTE:** The regional breakdown reflects geographic rather than economic or political groupings.

TABLE 2 (CONT.)

## Forest area and area change

| Country/area | Forest area, 2005 | | | | Annual change rate | | | |
|---|---|---|---|---|---|---|---|---|
| | Total forest[a] | % of land area | Area per capita | Forest plantations | 1990–2000 | | 2000–2005 | |
| | (1 000 ha) | (%) | (ha) | (1 000 ha) | (1 000 ha) | (%) | (1 000 ha) | (%) |
| Montserrat | 4 | 35 | 0.4 | – | 0 | 0 | 0 | 0 |
| Netherlands Antilles | 1 | 1.5 | 0.0 | – | 0 | 0 | 0 | 0 |
| Puerto Rico | 408 | 46 | 0.1 | – | n.s. | 0.1 | n.s. | n.s. |
| Saint Kitts and Nevis | 5 | 14.7 | 0.1 | – | 0 | 0 | 0 | 0 |
| Saint Lucia | 17 | 27.9 | 0.1 | – | 0 | 0 | 0 | 0 |
| Saint Vincent and the Grenadines | 11 | 27.4 | 0.1 | – | n.s. | 0.8 | n.s. | 0.8 |
| Trinidad and Tobago | 226 | 44.1 | 0.2 | 15 | –1 | –0.3 | n.s. | –0.2 |
| Turks and Caicos Islands | 34 | 80 | 1.6 | – | 0 | 0 | 0 | 0 |
| United States Virgin Islands | 10 | 27.9 | 0.1 | – | n.s. | –1.3 | n.s. | –1.8 |
| **Total Caribbean** | **5 974** | **26.1** | **0.2** | | **36** | **0.65** | **54** | **0.92** |
| | | | | | | | | |
| Belize | 1 653 | 72.5 | 5.8 | – | 0 | 0 | 0 | 0 |
| Costa Rica | 2 391 | 46.8 | 0.6 | 4 | –19 | –0.8 | 3 | 0.1 |
| El Salvador | 298 | 14.4 | 0.0 | – | –5 | –1.5 | –5 | –1.7 |
| Guatemala | 3 938 | 36.3 | 0.3 | – | –54 | –1.2 | –54 | –1.3 |
| Honduras | 4 648 | 41.5 | 0.7 | – | –196 | –3 | –156 | –3.1 |
| Nicaragua | 5 189 | 42.7 | 0.9 | – | –100 | –1.6 | –70 | –1.3 |
| Panama | 4 294 | 57.7 | 1.4 | 61 | –7 | –0.2 | –3 | –0.1 |
| **Total Central America** | **22 411** | **43.9** | **0.6** | | **–380** | **–1.47** | **–285** | **–1.23** |
| | | | | | | | | |
| Argentina | 33 021 | 12.1 | 0.9 | – | –149 | –0.4 | –150 | –0.4 |
| Bolivia | 58 740 | 54.2 | 6.5 | – | –270 | –0.4 | –270 | –0.5 |
| Brazil | 477 698 | 57.2 | 2.7 | – | –2 681 | –0.5 | –3 103 | –0.6 |
| Chile | 16 121 | 21.5 | 1.0 | 2 661 | 57 | 0.4 | 57 | 0.4 |
| Colombia | 60 728 | 58.5 | 1.3 | 328 | –48 | –0.1 | –47 | –0.1 |
| Ecuador | 10 853 | 39.2 | 0.8 | – | –198 | –1.5 | –198 | –1.7 |
| Falkland Islands | 0 | 0 | 0.0 | – | 0 | 0 | 0 | 0 |
| French Guiana | 8 063 | 91.8 | 41.1 | 1 | –3 | n.s. | 0 | 0 |
| Guyana | 15 104 | 76.7 | 19.6 | – | n.s. | n.s. | 0 | 0 |
| Paraguay | 18 475 | 46.5 | 3.2 | – | –179 | –0.9 | –179 | –0.9 |
| Peru | 68 742 | 53.7 | 2.5 | – | –94 | –0.1 | –94 | –0.1 |
| South Georgia and the South Sandwich Islands | 0 | 0 | 0.0 | – | 0 | 0 | 0 | 0 |
| Suriname | 14 776 | 94.7 | 33.4 | 7 | 0 | 0 | 0 | 0 |
| Uruguay | 1 506 | 8.6 | 0.4 | 766 | 50 | 4.5 | 19 | 1.3 |
| Venezuela (Bolivarian Republic of) | 47 713 | 54.1 | 1.8 | – | –288 | –0.6 | –288 | –0.6 |
| **Total South America** | **831 540** | **47.7** | **2.3** | | **–3 802** | **–0.44** | **–4 251** | **–0.5** |
| **Total Latin America and the Caribbean** | **859 925** | **47.3** | **1.9** | | **–4 147** | **–0.46** | **–4 483** | **–0.51** |

**NOTE:** The regional breakdown reflects geographic rather than economic or political groupings.

| Country/area | Forest area, 2005 | | | | Annual change rate | | | |
|---|---|---|---|---|---|---|---|---|
| | Total forest[a] | % of land area | Area per capita | Forest plantations | 1990–2000 | | 2000–2005 | |
| | (1 000 ha) | (%) | (ha) | (1 000 ha) | (1 000 ha) | (%) | (1 000 ha) | (%) |
| Canada | 310 134 | 33.6 | 9.7 | – | 0 | 0 | 0 | 0 |
| Greenland | n.s. | n.s. | n.s. | – | 0 | 0 | 0 | 0 |
| Mexico | 64 238 | 33.7 | 0.6 | 1 058 | –348 | –0.5 | –260 | –0.4 |
| Saint Pierre and Miquelon | 3 | 13 | 0.4 | – | 0 | 0 | 0 | 0 |
| United States of America | 303 089 | 33.1 | 1.0 | – | 365 | 0.1 | 159 | 0.1 |
| **Total North America** | **677 464** | **32.7** | **1.6** | | **17** | **n.s.** | **–101** | **–0.01** |
| | | | | | | | | |
| American Samoa | 18 | 89.4 | 0.3 | – | n.s. | –0.2 | n.s. | –0.2 |
| Australia | 163 678 | 21.3 | 8.1 | – | –326 | –0.2 | –193 | –0.1 |
| Cook Islands | 16 | 66.5 | 0.8 | – | n.s. | 0.4 | 0 | 0 |
| Fiji | 1 000 | 54.7 | 1.2 | – | 2 | 0.2 | 0 | 0 |
| French Polynesia | 105 | 28.7 | 0.4 | – | 0 | 0 | 0 | 0 |
| Guam | 26 | 47.1 | 0.2 | – | n.s. | n.s. | 0 | 0 |
| Kiribati | 2 | 3 | 0.0 | – | 0 | 0 | 0 | 0 |
| Marshall Islands | – | – | – | – | – | – | – | – |
| Micronesia (Federated States of) | 63 | 90.6 | 0.5 | – | 0 | 0 | 0 | 0 |
| Nauru | 0 | 0 | 0.0 | – | 0 | 0 | 0 | 0 |
| New Caledonia | 717 | 39.2 | 3.1 | – | 0 | 0 | 0 | 0 |
| New Zealand | 8 309 | 31 | 2.0 | 1 852 | 51 | 0.6 | 17 | 0.2 |
| Niue | 14 | 54.2 | 7.0 | – | n.s. | –1.3 | n.s. | –1.4 |
| Northern Mariana Islands | 33 | 72.4 | 0.4 | – | n.s. | –0.3 | n.s. | –0.3 |
| Palau | 40 | 87.6 | 2.0 | – | n.s. | 0.4 | n.s. | 0.4 |
| Papua New Guinea | 29 437 | 65 | 5.2 | – | –139 | –0.5 | –139 | –0.5 |
| Pitcairn | 4 | 83.3 | 80.0 | – | 0 | 0 | 0 | 0 |
| Samoa | 171 | 60.4 | 1.0 | 32 | 4 | 2.8 | 0 | 0 |
| Solomon Islands | 2 172 | 77.6 | 4.6 | – | –40 | –1.5 | –40 | –1.7 |
| Tokelau | 0 | 0 | 0.0 | – | 0 | 0 | 0 | 0 |
| Tonga | 4 | 5 | 0.0 | – | 0 | 0 | 0 | 0 |
| Tuvalu | 1 | 33.3 | 0.1 | – | 0 | 0 | 0 | 0 |
| Vanuatu | 440 | 36.1 | 2.0 | – | 0 | 0 | 0 | 0 |
| Wallis and Futuna Islands | 5 | 35.3 | 0.3 | 1 | n.s. | –0.8 | n.s. | –2 |
| **Total Oceania** | **206 254** | **24.3** | **6.3** | | **–448** | **–0.21** | **–356** | **–0.17** |
| | | | | | | | | |
| **Total World** | **3 952 025** | **30.3** | **0.6** | | **–8 868** | **–0.22** | **–7 317** | **–0.18** |

[a] "Total forest" includes forest plantations.
**SOURCE:** FAO, 2006a.

**NOTE:** The regional breakdown reflects geographic rather than economic or political groupings.

TABLE 3
## Forest growing stock, biomass and carbon

| Country/area | Growing stock | | | Biomass[a] | | Carbon in biomass[b] | |
|---|---|---|---|---|---|---|---|
| | Per hectare | Total | Commercial | Per hectare | Total | Per hectare | Total |
| | (m³/ha) | (million m³) | (% of total) | (tonnes/ha) | (million tonnes) | (tonnes/ha) | (million tonnes) |
| Burundi | – | – | – | – | – | – | – |
| Cameroon | 61.8 | 1 313 | 10.1 | 179.1 | 3 804 | 90 | 1 902 |
| Central African Republic | 167.0 | 3 801 | – | 246.3 | 5 604 | 123 | 2 801 |
| Chad | 18.3 | 218 | 38.1 | 39.5 | 471 | 20 | 236 |
| Congo | 202.5 | 4 551 | 30 | 461.1 | 10 361 | 231 | 5 181 |
| Democratic Republic of the Congo | 230.8 | 30 833 | – | 346.9 | 46 346 | 173 | 23 173 |
| Equatorial Guinea | 65.6 | 107 | – | 141.5 | 231 | 70 | 115 |
| Gabon | 222.5 | 4 845 | – | 334.6 | 7 285 | 167 | 3 643 |
| Rwanda | 183.3 | 88 | 95.1 | 183.3 | 88 | 92 | 44 |
| Saint Helena | – | – | – | – | – | – | – |
| Sao Tome and Principe | 148 | 4 | 100 | 333.3 | 9 | 148 | 4 |
| **Total Central Africa** | **194.0** | **45 760** | | **314.5** | **74 199** | **157.3** | **37 099** |
| | | | | | | | |
| British Indian Ocean Territory | – | – | – | – | – | – | – |
| Comoros | 200.0 | 1 | 26.9 | – | – | – | – |
| Djibouti | – | n.s. | – | – | – | – | – |
| Eritrea | – | – | – | – | – | – | – |
| Ethiopia | 21.9 | 285 | 25 | 38.7 | 503 | 19 | 252 |
| Kenya | 79.8 | 281 | 10.8 | 189.9 | 669 | 95 | 334 |
| Madagascar | 171.4 | 2 201 | 28.3 | 487.5 | 6 259 | 244 | 3 130 |
| Mauritius | 81.1 | 3 | 68 | 216.2 | 8 | 108 | 4 |
| Mayotte | – | – | – | – | – | – | – |
| Réunion | – | – | – | – | – | – | – |
| Seychelles | 75.0 | 3 | 12 | 175.0 | 7 | 100 | 4 |
| Somalia | 22.0 | 157 | n.s. | 108.5 | 774 | 54 | 387 |
| Uganda | 43.0 | 156 | 15 | 76.4 | 277 | 38 | 138 |
| United Republic of Tanzania | 35.9 | 1 264 | 73.3 | 127.9 | 4 509 | 64 | 2 254 |
| **Total East Africa** | **57.7** | **4 351** | | **172.4** | **13 006** | **86.2** | **6 503** |
| | | | | | | | |
| Algeria | 76.4 | 174 | 22 | 99.7 | 227 | 50 | 114 |
| Egypt | 119.4 | 8 | – | 209.0 | 14 | 104 | 7 |
| Libyan Arab Jamahiriya | 36.9 | 8 | – | 59.9 | 13 | 28 | 6 |
| Mauritania | 18.7 | 5 | – | 48.7 | 13 | 26 | 7 |
| Morocco | 43.8 | 191 | 100 | 110.0 | 480 | 55 | 240 |
| Sudan | 13.9 | 939 | – | 45.3 | 3 062 | 23 | 1 530 |
| Tunisia | 25.6 | 27 | 2.3 | 18.9 | 20 | 9 | 10 |
| Western Sahara | 37.6 | 38 | – | 50.4 | 51 | 25 | 25 |
| **Total Northern Africa** | **18.1** | **1 390** | | **50.5** | **3 880** | **25** | **1 939** |
| | | | | | | | |
| Angola | 38.8 | 2 291 | 1.1 | 163.4 | 9 658 | 82 | 4 830 |
| Botswana | 16.5 | 197 | – | 23.7 | 283 | 12 | 141 |
| Lesotho | – | – | – | – | – | – | – |

n.s. = not significant, indicating a very small value
– = not available
0 = true zero
**NOTE:** The regional breakdown reflects geographic rather than economic or political groupings.

| Country/area | Growing stock | | | Biomass[a] | | Carbon in biomass[b] | |
|---|---|---|---|---|---|---|---|
| | Per hectare | Total | Commercial | Per hectare | Total | Per hectare | Total |
| | (m³/ha) | (million m³) | (% of total) | (tonnes/ha) | (million tonnes) | (tonnes/ha) | (million tonnes) |
| Malawi | 109.6 | 373 | – | 94.7 | 322 | 47 | 161 |
| Mozambique | 25.8 | 496 | 14.4 | 63.0 | 1 213 | 31 | 606 |
| Namibia | 24.0 | 184 | – | 60.3 | 462 | 30 | 231 |
| South Africa | 69.0 | 635 | 38.1 | 179.1 | 1 648 | 90 | 824 |
| Swaziland | 35.1 | 19 | – | 86.9 | 47 | 43 | 23 |
| Zambia | 30.8 | 1 307 | 7.1 | 54.5 | 2 313 | 27 | 1 156 |
| Zimbabwe | 34.2 | 600 | 3.8 | 60.9 | 1 069 | 31 | 535 |
| **Total Southern Africa** | **35.7** | **6 102** | | **99.4** | **17 015** | **49.7** | **8 507** |
| | | | | | | | |
| Benin | – | – | – | – | – | – | – |
| Burkina Faso | 35.0 | 238 | 4.6 | 87.7 | 596 | 44 | 298 |
| Cape Verde | 142.9 | 12 | 80 | 190.5 | 16 | 95 | 8 |
| Côte d'Ivoire | 257.9 | 2 683 | 19.9 | 385.8 | 4 014 | 179 | 1 864 |
| Gambia | 38.2 | 18 | – | 140.1 | 66 | 70 | 33 |
| Ghana | 58.2 | 321 | 53.3 | 180.0 | 993 | 90 | 496 |
| Guinea | 77.3 | 520 | – | 189.2 | 1 272 | 95 | 636 |
| Guinea-Bissau | 24.1 | 50 | 20 | 58.9 | 122 | 29 | 61 |
| Liberia | 157.9 | 498 | 41.5 | 287.3 | 906 | 144 | 453 |
| Mali | 15.2 | 191 | – | 38.5 | 484 | 19 | 242 |
| Niger | 10.3 | 13 | 8.1 | 19.7 | 25 | 9 | 12 |
| Nigeria | 125.0 | 1 386 | 10.9 | 252.9 | 2 804 | 126 | 1 401 |
| Senegal | 37.4 | 324 | 63.3 | 85.4 | 741 | 43 | 371 |
| Sierra Leone | – | – | – | – | – | – | – |
| Togo | – | – | – | – | – | – | – |
| **Total West Africa** | **90.9** | **6 254** | | **174.9** | **12 039** | **85.4** | **5 875** |
| **Total Africa** | **101.7** | **63 857** | | **191.3** | **120 139** | **95.4** | **59 923** |
| | | | | | | | |
| Afghanistan | 16.1 | 14 | 40 | 15.0 | 13 | 7 | 6 |
| Armenia | 127.2 | 36 | – | 127.2 | 36 | 64 | 18 |
| Azerbaijan | 135.7 | 127 | 20.4 | 123.9 | 116 | 62 | 58 |
| Georgia | 167.0 | 461 | 26.2 | 152.2 | 420 | 76 | 210 |
| Kazakhstan | 109.1 | 364 | 0 | 82.1 | 274 | 41 | 136 |
| Kyrgyzstan | 34.5 | 30 | 0 | 28.8 | 25 | 14 | 12 |
| Tajikistan | 12.2 | 5 | 0 | 14.6 | 6 | 7 | 3 |
| Turkmenistan | 3.4 | 14 | 0 | 8.2 | 34 | 4 | 18 |
| Uzbekistan | 7.3 | 24 | 0.1 | 7.3 | 24 | 4 | 13 |
| **Total Central Asia** | **63.7** | **1 075** | | **56.1** | **948** | **28.1** | **474** |
| | | | | | | | |
| China | 67.2 | 13 255 | 91.8 | 61.8 | 12 191 | 31 | 6 096 |
| Democratic People's Republic of Korea | 63.8 | 395 | – | 75.2 | 465 | 37 | 232 |
| Japan | 170.9 | 4 249 | – | 152.2 | 3 785 | 76 | 1 892 |
| Mongolia | 130.9 | 1 342 | 46.1 | 112.0 | 1 148 | 56 | 574 |
| Republic of Korea | 80.1 | 502 | 53.6 | 82.2 | 515 | 41 | 258 |
| **Total East Asia** | **80.6** | **19 743** | | **73.9** | **18 104** | **37.0** | **9 052** |

**NOTE:** The regional breakdown reflects geographic rather than economic or political groupings.

TABLE 3 (CONT.)

**Forest growing stock, biomass and carbon**

| Country/area | Growing stock | | | Biomass[a] | | Carbon in biomass[b] | |
|---|---|---|---|---|---|---|---|
| | Per hectare | Total | Commercial | Per hectare | Total | Per hectare | Total |
| | (m³/ha) | (million m³) | (% of total) | (tonnes/ha) | (million tonnes) | (tonnes/ha) | (million tonnes) |
| Bangladesh | 34.4 | 30 | 75 | 72.3 | 63 | 36 | 31 |
| Bhutan | 194.4 | 621 | 40.1 | 216.0 | 690 | 108 | 345 |
| India | 69.4 | 4 698 | 40 | 76.5 | 5 178 | 35 | 2 343 |
| Maldives | – | – | – | – | – | – | – |
| Nepal | 177.9 | 647 | 40 | 266.5 | 969 | 133 | 485 |
| Pakistan | 97.3 | 185 | 43.2 | 271.3 | 516 | 136 | 259 |
| Sri Lanka | 21.7 | 42 | 40 | 40.9 | 79 | 21 | 40 |
| **Total South Asia** | **78.5** | **6 223** | | **94.6** | **7 495** | **44.2** | **3 503** |
| | | | | | | | |
| Brunei Darussalam | 219.4 | 61 | 40 | 280.6 | 78 | 144 | 40 |
| Cambodia | 95.5 | 998 | 40 | 242.4 | 2 532 | 121 | 1 266 |
| Indonesia | 58.9 | 5 216 | – | 133.3 | 11 793 | 67 | 5 897 |
| Lao People's Democratic Republic | 59.3 | 957 | 74 | 184.2 | 2 974 | 92 | 1 487 |
| Malaysia | 250.9 | 5 242 | – | 336.0 | 7 020 | 168 | 3 510 |
| Myanmar | 85.0 | 2 740 | 17.8 | 196.6 | 6 335 | 98 | 3 168 |
| Philippines | 174.3 | 1 248 | 4.3 | 271.2 | 1 942 | 136 | 971 |
| Singapore | – | – | – | – | – | – | – |
| Thailand | 41.3 | 599 | 59.9 | 98.8 | 1 434 | 49 | 716 |
| Timor-Leste | – | – | – | – | – | – | – |
| Viet Nam | 65.7 | 850 | 8.5 | 181.6 | 2 348 | 91 | 1 174 |
| **Total Southeast Asia** | **88.2** | **17 911** | | **179.5** | **36 456** | **89.8** | **18 229** |
| | | | | | | | |
| Bahrain | – | – | – | – | – | – | – |
| Cyprus | 46.0 | 8 | 39 | 28.7 | 5 | 17 | 3 |
| Iran (Islamic Republic of) | 47.6 | 527 | 78.9 | 60.4 | 669 | 30 | 334 |
| Iraq | – | – | – | – | – | – | – |
| Israel | 35.1 | 6 | 70 | – | – | – | – |
| Jordan | 24.1 | 2 | – | 48.2 | 4 | 36 | 3 |
| Kuwait | – | – | – | – | – | – | – |
| Lebanon | 36.8 | 5 | – | 29.4 | 4 | 15 | 2 |
| Occupied Palestinian Territory | – | – | – | – | – | – | – |
| Oman | – | – | – | – | – | – | – |
| Qatar | – | – | – | – | – | – | – |
| Saudi Arabia | 8.4 | 23 | 0 | 12.8 | 35 | 7 | 18 |
| Syrian Arab Republic | – | – | – | – | – | – | – |
| Turkey | 137.6 | 1 400 | 86.6 | 160.5 | 1 633 | 80 | 817 |
| United Arab Emirates | 48.1 | 15 | 0 | 105.8 | 33 | 54 | 17 |
| Yemen | 9.1 | 5 | – | 18.2 | 10 | 11 | 6 |
| **Total Western Asia** | **78.4** | **1 991** | | **94.2** | **2 393** | **47.2** | **1 200** |
| **Total Asia** | **82.4** | **46 943** | | **114.9** | **65 396** | **57.0** | **32 458** |

**NOTE:** The regional breakdown reflects geographic rather than economic or political groupings.

| Country/area | Growing stock | | | Biomass[a] | | Carbon in biomass[b] | |
|---|---|---|---|---|---|---|---|
| | Per hectare | Total | Commercial | Per hectare | Total | Per hectare | Total |
| | *(m³/ha)* | *(million m³)* | *(% of total)* | *(tonnes/ha)* | *(million tonnes)* | *(tonnes/ha)* | *(million tonnes)* |
| Albania | 98.2 | 78 | 81 | 129.7 | 103 | 65 | 52 |
| Andorra | – | – | – | – | – | – | – |
| Austria | 300.1 | 1 159 | 97.7 | – | – | – | – |
| Belarus | 178.7 | 1 411 | 82.8 | 136.7 | 1 079 | 68 | 539 |
| Belgium | 257.9 | 172 | 100 | 194.9 | 130 | 97 | 65 |
| Bosnia and Herzegovina | 178.9 | 391 | 80.1 | 160.6 | 351 | 81 | 176 |
| Bulgaria | 156.7 | 568 | 61.1 | 145.4 | 527 | 73 | 263 |
| Channel Islands | – | – | – | – | – | – | – |
| Croatia | 164.9 | 352 | 83 | 179.9 | 384 | 90 | 192 |
| Czech Republic | 277.9 | 736 | 96.7 | 273.8 | 725 | 123 | 327 |
| Denmark | 152.0 | 76 | 76.1 | 104.0 | 52 | 52 | 26 |
| Estonia | 195.7 | 447 | 93.7 | 146.2 | 334 | 74 | 168 |
| Faeroe Islands | – | – | – | – | – | – | – |
| Finland | 95.9 | 2 158 | 84.1 | 72.5 | 1 632 | 36 | 815 |
| France | 158.5 | 2 465 | 93.5 | 157.6 | 2 452 | 75 | 1 165 |
| Germany | – | – | – | 235.2 | 2 605 | 118 | 1 303 |
| Gibraltar | – | – | – | – | – | – | – |
| Greece | 47.2 | 177 | 88.1 | 31.2 | 117 | 16 | 59 |
| Holy See | – | – | – | – | – | – | – |
| Hungary | 170.5 | 337 | 97.6 | 172.1 | 340 | 88 | 173 |
| Iceland | 65.2 | 3 | – | – | – | – | – |
| Ireland | 97.2 | 65 | – | 59.8 | 40 | 28 | 19 |
| Isle of Man | – | – | – | – | – | – | – |
| Italy | 145.0 | 1 447 | 70.1 | 127.6 | 1 273 | 64 | 636 |
| Latvia | 203.7 | 599 | 85.3 | 157.1 | 462 | 78 | 230 |
| Liechtenstein | 285.7 | 2 | 80 | – | – | – | – |
| Lithuania | 190.6 | 400 | 86 | 122.9 | 258 | 61 | 129 |
| Luxembourg | 298.9 | 26 | 100 | 218.4 | 19 | 103 | 9 |
| Malta | – | n.s. | 0 | – | – | – | – |
| Monaco | – | – | – | – | – | – | – |
| Netherlands | 178.1 | 65 | 80 | 142.5 | 52 | 68 | 25 |
| Norway | 91.9 | 863 | 78.2 | 73.5 | 690 | 37 | 344 |
| Poland | 202.8 | 1 864 | 94.4 | 194.8 | 1 791 | 97 | 896 |
| Portugal | 92.5 | 350 | 66.3 | 60.3 | 228 | 30 | 114 |
| Republic of Moldova | 142.9 | 47 | 62.3 | 79.0 | 26 | 40 | 13 |
| Romania | 211.5 | 1 347 | 98 | 177.9 | 1 133 | 89 | 567 |
| Russian Federation | 99.5 | 80 479 | 49.2 | 79.6 | 64 420 | 40 | 32 210 |
| San Marino | – | – | – | – | – | – | – |
| Serbia and Montenegro | 121.4 | 327 | – | 115.8 | 312 | 58 | 156 |
| Slovakia | 256.1 | 494 | 84.7 | 211.0 | 407 | 105 | 203 |
| Slovenia | 282.4 | 357 | 91.3 | 232.6 | 294 | 116 | 147 |
| Spain | 49.6 | 888 | 77.6 | 48.6 | 871 | 22 | 392 |

**NOTE:** The regional breakdown reflects geographic rather than economic or political groupings.

TABLE 3 (CONT.)

## Forest growing stock, biomass and carbon

| Country/area | Growing stock | | | Biomass[a] | | Carbon in biomass[b] | |
|---|---|---|---|---|---|---|---|
| | Per hectare | Total | Commercial | Per hectare | Total | Per hectare | Total |
| | *(m³/ha)* | *(million m³)* | *(% of total)* | *(tonnes/ha)* | *(million tonnes)* | *(tonnes/ha)* | *(million tonnes)* |
| Sweden | 114.6 | 3 155 | 76.8 | 85.0 | 2 340 | 43 | 1 170 |
| Switzerland | 367.7 | 449 | 82.4 | 252.3 | 308 | 126 | 154 |
| The former Yugoslav Republic of Macedonia | 69.5 | 63 | – | 45.3 | 41 | 22 | 20 |
| Ukraine | 221.3 | 2 119 | 63.8 | 155.5 | 1 489 | 78 | 745 |
| United Kingdom | 119.5 | 340 | 88.2 | 78.7 | 224 | 39 | 112 |
| **Total Europe** | **107.3** | **106 276** | | **87.7** | **87 509** | **43.7** | **43 614** |
| | | | | | | | |
| Anguilla | – | – | – | – | – | – | – |
| Antigua and Barbuda | – | – | – | – | – | – | – |
| Aruba | – | – | – | – | – | – | – |
| Bahamas | 13.6 | 7 | – | – | – | – | – |
| Barbados | – | – | – | – | – | – | – |
| Bermuda | – | – | – | – | – | – | – |
| British Virgin Islands | – | – | – | – | – | – | – |
| Cayman Islands | – | – | – | – | – | – | – |
| Cuba | 89.6 | 243 | 78.6 | 272.8 | 740 | 128 | 347 |
| Dominica | – | – | – | – | – | – | – |
| Dominican Republic | 46.5 | 64 | – | 119.2 | 164 | 60 | 82 |
| Grenada | – | – | – | – | – | – | – |
| Guadeloupe | – | – | – | – | – | – | – |
| Haiti | 66.7 | 7 | – | 161.9 | 17 | 76 | 8 |
| Jamaica | 156.3 | 53 | 2.1 | 200.6 | 68 | 100 | 34 |
| Martinique | – | – | – | – | – | – | – |
| Montserrat | – | – | – | – | – | – | – |
| Netherlands Antilles | – | – | – | – | – | – | – |
| Puerto Rico | 63.7 | 26 | – | 102.9 | 42 | 51 | 21 |
| Saint Kitts and Nevis | – | – | – | – | – | – | – |
| Saint Lucia | – | – | – | – | – | – | – |
| Saint Vincent and the Grenadines | – | – | – | – | – | – | – |
| Trinidad and Tobago | 88.5 | 20 | 55 | 208.0 | 47 | 106 | 24 |
| Turks and Caicos Islands | – | – | – | – | – | – | – |
| United States Virgin Islands | – | n.s. | – | – | – | – | – |
| **Total Caribbean** | **73.8** | **420** | | **208.2** | **1 078** | **99.7** | **516** |
| | | | | | | | |
| Belize | 96.2 | 159 | – | 71.4 | 118 | 36 | 59 |
| Costa Rica | 104.1 | 249 | 66.3 | 161.0 | 385 | 81 | 193 |
| El Salvador | – | – | – | – | – | – | – |
| Guatemala | 163.0 | 642 | 15.5 | 252.9 | 996 | 126 | 498 |
| Honduras | 116.2 | 540 | – | – | – | – | – |
| Nicaragua | 113.9 | 591 | 24.9 | 276.0 | 1 432 | 138 | 716 |
| Panama | 159.8 | 686 | 1.3 | 288.3 | 1 238 | 144 | 620 |
| **Total Central America** | **129.7** | **2 867** | | **238.7** | **4 169** | **119.4** | **2 086** |

**NOTE:** The regional breakdown reflects geographic rather than economic or political groupings.

| Country/area | Growing stock | | | Biomass[a] | | Carbon in biomass[b] | |
|---|---|---|---|---|---|---|---|
| | Per hectare | Total | Commercial | Per hectare | Total | Per hectare | Total |
| | *(m³/ha)* | *(million m³)* | *(% of total)* | *(tonnes/ha)* | *(million tonnes)* | *(tonnes/ha)* | *(million tonnes)* |
| Argentina | 55.3 | 1 826 | 67.1 | 145.9 | 4 817 | 73 | 2 411 |
| Bolivia | 74.2 | 4 360 | 15.5 | 179.9 | 10 568 | 90 | 5 296 |
| Brazil | 170.1 | 81 239 | 18.1 | 211.9 | 101 236 | 103 | 49 335 |
| Chile | 116.7 | 1 882 | 64.3 | 241.4 | 3 892 | 121 | 1 946 |
| Colombia | – | – | – | 265.5 | 16 125 | 133 | 8 062 |
| Ecuador | – | – | – | – | – | – | – |
| Falkland Islands | – | – | – | – | – | – | – |
| French Guiana | 350.0 | 2 822 | 0.3 | – | – | – | – |
| Guyana | – | – | – | 228.0 | 3 443 | 114 | 1 722 |
| Paraguay | – | – | – | – | – | – | – |
| Peru | – | – | – | – | – | – | – |
| South Georgia and the South Sandwich Islands | – | – | – | – | – | – | – |
| Suriname | 150.0 | 2 216 | – | 770.4 | 11 383 | 385 | 5 692 |
| Uruguay | 78.4 | 118 | 6.2 | – | – | – | – |
| Venezuela (Bolivarian Republic of) | – | – | – | – | – | – | – |
| **Total South America** | **154.9** | **94 463** | | **224.0** | **151 464** | **110.1** | **74 464** |
| **Total Latin America and the Caribbean** | **153.3** | **97 750** | | **224.2** | **156 711** | **110.3** | **77 066** |
| | | | | | | | |
| Canada | 106.4 | 32 983 | 100 | – | – | – | – |
| Greenland | – | – | – | – | – | – | – |
| Mexico | – | – | – | – | – | – | – |
| Saint Pierre and Miquelon | – | – | – | – | – | – | – |
| United States of America | 115.9 | 35 118 | 78.7 | 125.1 | 37 929 | 63 | 18 964 |
| **Total North America** | **111.1** | **68 101** | | **125.1** | **37 929** | **62.6** | **18 964** |
| | | | | | | | |
| American Samoa | 111.1 | 2 | – | 222.2 | 4 | 111 | 2 |
| Australia | – | – | – | 113.1 | 18 510 | 51 | 8 339 |
| Cook Islands | – | – | – | – | – | – | – |
| Fiji | – | – | – | – | – | – | – |
| French Polynesia | – | – | – | – | – | – | – |
| Guam | – | – | – | – | – | – | – |
| Kiribati | – | – | – | – | – | – | – |
| Marshall Islands | – | – | – | – | – | – | – |
| Micronesia (Federated States of) | – | – | – | – | – | – | – |
| Nauru | – | – | – | – | – | – | – |
| New Caledonia | 55.8 | 40 | 58.2 | 203.6 | 146 | 102 | 73 |
| New Zealand | – | – | – | – | – | – | – |
| Niue | – | – | – | – | – | – | – |
| Northern Mariana Islands | – | – | – | – | – | – | – |
| Palau | – | – | – | – | – | – | – |
| Papua New Guinea | 35.2 | 1 035 | 50.7 | – | – | – | – |

**NOTE:** The regional breakdown reflects geographic rather than economic or political groupings.

TABLE 3 (CONT.)

## Forest growing stock, biomass and carbon

| Country/area | Growing stock | | | Biomass[a] | | Carbon in biomass[b] | |
|---|---|---|---|---|---|---|---|
| | Per hectare | Total | Commercial | Per hectare | Total | Per hectare | Total |
| | *($m^3$/ha)* | *(million $m^3$)* | *(% of total)* | *(tonnes/ha)* | *(million tonnes)* | *(tonnes/ha)* | *(million tonnes)* |
| Pitcairn | – | – | – | – | – | – | – |
| Samoa | – | – | – | – | – | – | – |
| Solomon Islands | – | – | – | – | – | – | – |
| Tokelau | – | – | – | – | – | – | – |
| Tonga | – | – | – | – | – | – | – |
| Tuvalu | – | – | – | – | – | – | – |
| Vanuatu | – | – | – | – | – | – | – |
| Wallis and Futuna Islands | – | – | – | – | – | – | – |
| **Total Oceania** | **35.7** | **1 077** | | **113.5** | **18 660** | **51.2** | **8 414** |
| **Total World** | **110.7** | **384 004** | | **144.7** | **486 344** | **71.5** | **240 439** |

[a] "Biomass" includes above- and below-ground biomass and dead wood.
[b] "Carbon in biomass" excludes carbon in dead wood, litter and soil.
**SOURCE:** FAO, 2006a.

**NOTE:** The regional breakdown reflects geographic rather than economic or political groupings.

TABLE 4

## Production, trade and consumption of roundwood and sawnwood, 2004

| Country/area | Woodfuel (1 000 m³) | | | | Industrial roundwood (1 000 m³) | | | | Sawnwood (1 000 m³) | | | |
|---|---|---|---|---|---|---|---|---|---|---|---|---|
| | Production | Imports | Exports | Consumption | Production | Imports | Exports | Consumption | Production | Imports | Exports | Consumption |
| Burundi | 8 390 | 0 | 0 | 8 390 | 333 | 0 | 7 | 326 | 83 | 0 | 0 | 83 |
| Cameroon | 9 407 | 0 | 0 | 9 407 | 1 800 | 0 | 29 | 1 771 | 702 | 0 | 514 | 188 |
| Central African Republic | 2 000 | 0 | 0 | 2 000 | 832 | 1 | 364 | 469 | 69 | 0 | 20 | 49 |
| Chad | 6 362 | 0 | 0 | 6 362 | 761 | 0 | 0 | 761 | 2 | 17 | 1 | 19 |
| Congo | 1 219 | 0 | 0 | 1 219 | 896 | 0 | 844 | 52 | 157 | 0 | 143 | 14 |
| Democratic Republic of the Congo | 69 777 | 0 | 0 | 69 777 | 3 653 | 0 | 236 | 3 417 | 15 | 0 | 14 | 1 |
| Equatorial Guinea | 447 | 0 | 0 | 447 | 700 | 0 | 685 | 15 | 4 | 0 | 1 | 3 |
| Gabon | 1 070 | 0 | 0 | 1 070 | 3 500 | 0 | 1 718 | 1 782 | 133 | 0 | 81 | 52 |
| Rwanda | 5 000 | 0 | 0 | 5 000 | 495 | 0 | 0 | 495 | 79 | 0 | 0 | 79 |
| Saint Helena | – | – | – | – | 0 | 0 | 0 | 0 | 0 | 0 | 0 | 0 |
| Sao Tome and Principe | 0 | 0 | 0 | 0 | 9 | 0 | 0 | 9 | 5 | 0 | 1 | 5 |
| **Total Central Africa** | **103 673** | **0** | **0** | **103 673** | **12 979** | **2** | **3 883** | **9 097** | **1 250** | **19** | **775** | **494** |
| British Indian Ocean Territory | – | – | – | – | – | – | – | – | – | – | – | – |
| Comoros | 0 | 0 | 0 | 0 | 9 | 0 | 0 | 9 | 0 | 1 | 0 | 1 |
| Djibouti | 0 | 0 | 0 | 0 | 0 | 1 | 0 | 1 | 0 | 2 | 0 | 2 |
| Eritrea | 2 406 | 0 | 0 | 2 406 | 2 | 4 | 0 | 6 | 0 | 0 | 0 | 0 |
| Ethiopia | 93 029 | 0 | 0 | 93 029 | 2 928 | 0 | 0 | 2 928 | 18 | 10 | 0 | 28 |
| Kenya | 20 370 | 0 | 0 | 20 370 | 1 792 | 2 | 5 | 1 788 | 78 | 5 | 1 | 82 |
| Madagascar | 10 770 | 0 | 0 | 10 770 | 183 | 0 | 43 | 140 | 893 | 1 | 28 | 866 |
| Mauritius | 6 | 0 | 0 | 6 | 8 | 20 | 1 | 27 | 3 | 65 | 1 | 67 |
| Mayotte | – | – | – | – | – | – | – | – | – | – | – | – |
| Réunion | 31 | 0 | 0 | 31 | 5 | 1 | 2 | 3 | 2 | 85 | 0 | 87 |
| Seychelles | – | – | – | – | 0 | 0 | 0 | 0 | 0 | 0 | 0 | 0 |
| Somalia | 10 466 | 0 | 0 | 10 466 | 110 | 1 | 5 | 106 | 14 | 0 | 0 | 14 |
| Uganda | 36 235 | 0 | 0 | 36 235 | 3 175 | 0 | 4 | 3 171 | 264 | 0 | 1 | 264 |
| United Republic of Tanzania | 21 505 | 0 | 2 | 21 503 | 2 314 | 6 | 74 | 2 246 | 24 | 0 | 12 | 12 |
| **Total East Africa** | **194 818** | **0** | **2** | **194 816** | **10 526** | **35** | **135** | **10 426** | **1 296** | **168** | **43** | **1 422** |
| Algeria | 7 545 | 0 | 0 | 7 545 | 119 | 79 | 1 | 197 | 13 | 1 329 | 1 | 1 341 |
| Egypt | 16 792 | 0 | 0 | 16 792 | 268 | 116 | 0 | 384 | 2 | 1 463 | 0 | 1 465 |
| Libyan Arab Jamahiriya | 536 | 0 | 0 | 536 | 116 | 8 | 0 | 124 | 31 | 123 | 0 | 154 |
| Mauritania | 1 581 | 0 | 0 | 1 581 | 6 | 1 | 0 | 7 | 0 | 2 | 0 | 1 |
| Morocco | 298 | 0 | 0 | 298 | 563 | 653 | 1 | 1 215 | 83 | 1 051 | 1 | 1 133 |
| Sudan | 17 482 | 0 | 0 | 17 482 | 2 173 | 0 | 2 | 2 171 | 51 | 77 | 1 | 126 |
| Tunisia | 2 138 | 0 | 0 | 2 138 | 214 | 81 | 0 | 295 | 20 | 562 | 2 | 581 |

0 = either a true zero or an insignificant value (less than half a unit)
**NOTE:** The regional breakdown reflects geographic rather than economic or political groupings.

TABLE 4 (CONT.)

## Production, trade and consumption of roundwood and sawnwood, 2004

| Country/area | Woodfuel (1 000 m³) | | | | Industrial roundwood (1 000 m³) | | | | Sawnwood (1 000 m³) | | | |
|---|---|---|---|---|---|---|---|---|---|---|---|---|
| | Production | Imports | Exports | Consumption | Production | Imports | Exports | Consumption | Production | Imports | Exports | Consumption |
| Western Sahara | – | – | – | – | – | – | – | – | – | – | – | – |
| **Total Northern Africa** | **46 371** | **0** | **0** | **46 371** | **3 458** | **938** | **4** | **4 393** | **200** | **4 607** | **5** | **4 802** |
| Angola | 3 487 | 0 | 0 | 3 487 | 1 096 | 3 | 12 | 1 087 | 5 | 3 | 0 | 8 |
| Botswana | 655 | 0 | 0 | 655 | 105 | 0 | 0 | 105 | 0 | 15 | 0 | 15 |
| Lesotho | 2 047 | 0 | 0 | 2 047 | – | – | – | – | 0 | 0 | 0 | 0 |
| Malawi | 5 102 | 0 | 0 | 5 102 | 520 | 1 | 0 | 520 | 45 | 0 | 1 | 44 |
| Mozambique | 16 724 | 0 | 0 | 16 724 | 1 319 | 16 | 89 | 1 246 | 28 | 13 | 17 | 24 |
| Namibia | – | – | – | – | 0 | 0 | 0 | 0 | 0 | 0 | 0 | 0 |
| South Africa | 12 000 | 0 | 0 | 12 000 | 21 159 | 38 | 371 | 20 827 | 2 171 | 360 | 84 | 2 447 |
| Swaziland | 560 | 0 | 0 | 560 | 330 | 0 | 0 | 330 | 102 | 0 | 0 | 102 |
| Zambia | 7 219 | 0 | 0 | 7 219 | 834 | 0 | 1 | 834 | 157 | 1 | 6 | 153 |
| Zimbabwe | 8 115 | 0 | 0 | 8 115 | 992 | 1 | 5 | 989 | 397 | 2 | 83 | 317 |
| **Total Southern Africa** | **55 908** | **0** | **0** | **55 908** | **26 356** | **60** | **478** | **25 937** | **2 905** | **396** | **191** | **3 110** |
| Benin | 162 | 0 | 0 | 162 | 332 | 15 | 34 | 313 | 31 | 0 | 5 | 25 |
| Burkina Faso | 8 040 | 0 | 0 | 8 040 | 1 171 | 3 | 3 | 1 171 | 1 | 21 | 4 | 18 |
| Cape Verde | 2 | 0 | 0 | 2 | 0 | 3 | 0 | 2 | 0 | 7 | 0 | 7 |
| Côte d'Ivoire | 8 655 | 0 | 0 | 8 655 | 1 678 | 10 | 120 | 1 568 | 512 | 0 | 359 | 153 |
| Gambia | 638 | 0 | 0 | 638 | 113 | 1 | 2 | 111 | 1 | 1 | 0 | 1 |
| Ghana | 20 678 | 0 | 0 | 20 678 | 1 350 | 3 | 1 | 1 351 | 480 | 0 | 211 | 270 |
| Guinea | 11 635 | 0 | 0 | 11 635 | 651 | 1 | 23 | 629 | 26 | 0 | 9 | 18 |
| Guinea-Bissau | 422 | 0 | 0 | 422 | 170 | 0 | 7 | 163 | 16 | 1 | 0 | 16 |
| Liberia | 5 576 | 0 | 0 | 5 576 | 250 | 0 | 0 | 250 | 20 | 2 | 1 | 20 |
| Mali | 4 965 | 0 | 0 | 4 965 | 413 | 1 | 1 | 413 | 13 | 0 | 0 | 13 |
| Niger | 8 596 | 0 | 0 | 8 596 | 411 | 1 | 4 | 408 | 4 | 0 | 0 | 4 |
| Nigeria | 60 852 | 0 | 1 | 60 851 | 9 418 | 1 | 42 | 9 377 | 2 000 | 1 | 22 | 1 980 |
| Senegal | 5 243 | 0 | 0 | 5 243 | 794 | 23 | 0 | 817 | 23 | 86 | 1 | 108 |
| Sierra Leone | 5 403 | 0 | 0 | 5 403 | 124 | 0 | 1 | 122 | 5 | 2 | 1 | 6 |
| Togo | 4 424 | 0 | 0 | 4 424 | 254 | 1 | 5 | 250 | 13 | 1 | 0 | 14 |
| **Total West Africa** | **145 292** | **0** | **1** | **145 291** | **17 128** | **62** | **245** | **16 945** | **3 145** | **122** | **613** | **2 653** |
| **Total Africa** | **546 062** | **1** | **4** | **546 059** | **70 447** | **1 097** | **4 745** | **66 799** | **8 796** | **5 312** | **1 627** | **12 480** |
| Afghanistan | 1 427 | 0 | 0 | 1 427 | 1 760 | 6 | 100 | 1 666 | 400 | 123 | 0 | 522 |
| Armenia | 59 | 0 | 0 | 59 | 6 | 1 | 1 | 7 | 2 | 28 | 8 | 22 |
| Azerbaijan | 6 | 0 | 0 | 6 | 7 | 6 | 0 | 14 | 0 | 302 | 2 | 300 |
| Georgia | 443 | 0 | 1 | 442 | 100 | 0 | 56 | 43 | 69 | 3 | 48 | 25 |
| Kazakhstan | 171 | 1 | 0 | 172 | 130 | 156 | 0 | 286 | 265 | 539 | 126 | 678 |
| Kyrgyzstan | 18 | 4 | 0 | 22 | 9 | 4 | 0 | 13 | 22 | 0 | 1 | 22 |
| Tajikistan | 0 | 0 | 0 | 0 | 0 | 0 | 0 | 0 | 0 | 40 | 0 | 40 |

**NOTE:** The regional breakdown reflects geographic rather than economic or political groupings.

| Country/area | Woodfuel (1 000 m³) | | | | Industrial roundwood (1 000 m³) | | | | Sawnwood (1 000 m³) | | | |
|---|---|---|---|---|---|---|---|---|---|---|---|---|
| | Production | Imports | Exports | Consumption | Production | Imports | Exports | Consumption | Production | Imports | Exports | Consumption |
| Turkmenistan | 3 | 0 | 0 | 3 | 0 | 0 | 0 | 0 | 0 | 24 | 0 | 24 |
| Uzbekistan | 18 | 0 | 1 | 17 | 8 | 234 | 1 | 241 | 0 | 6 | 0 | 6 |
| **Total Central Asia** | **2 146** | **5** | **2** | **2 148** | **2 020** | **407** | **158** | **2 269** | **759** | **1 064** | **186** | **1 637** |
| China | 191 044 | 7 | 6 | 191 045 | 95 061 | 27 642 | 710 | 121 993 | 12 211 | 7 628 | 701 | 19 138 |
| Democratic People's Republic of Korea | 5 737 | 0 | 0 | 5 737 | 1 500 | 0 | 40 | 1 460 | 280 | 1 | 22 | 259 |
| Japan | 114 | 1 | 0 | 115 | 15 615 | 12 681 | 9 | 28 287 | 13 603 | 9 123 | 18 | 22 708 |
| Mongolia | 186 | 0 | 0 | 186 | 445 | 7 | 1 | 451 | 300 | 2 | 3 | 299 |
| Republic of Korea | 2 463 | 0 | 0 | 2 463 | 2 089 | 6 540 | 0 | 8 629 | 4 366 | 834 | 17 | 5 183 |
| **Total East Asia** | **199 545** | **8** | **6** | **199 547** | **114 710** | **46 870** | **759** | **160 821** | **30 760** | **17 588** | **761** | **47 587** |
| Bangladesh | 27 694 | 0 | 0 | 27 694 | 282 | 344 | 1 | 625 | 388 | 4 | 4 | 388 |
| Bhutan | 4 479 | 0 | 0 | 4 479 | 133 | 0 | 3 | 130 | 31 | 0 | 0 | 31 |
| India | 303 839 | 0 | 0 | 303 839 | 19 146 | 1 933 | 9 | 21 069 | 17 500 | 54 | 20 | 17 534 |
| Maldives | 0 | 0 | 0 | 0 | 0 | 0 | 0 | 0 | 0 | 0 | 0 | 0 |
| Nepal | 12 702 | 0 | 0 | 12 702 | 1 260 | 1 | 1 | 1 260 | 630 | 2 | 0 | 631 |
| Pakistan | 25 599 | 0 | 0 | 25 599 | 2 679 | 202 | 0 | 2 881 | 1 180 | 91 | 0 | 1 271 |
| Sri Lanka | 5 646 | 0 | 0 | 5 646 | 694 | 1 | 15 | 680 | 61 | 44 | 1 | 104 |
| **Total South Asia** | **379 960** | **0** | **0** | **379 960** | **24 194** | **2 481** | **29** | **26 645** | **19 790** | **194** | **25** | **19 959** |
| Brunei Darussalam | 12 | 0 | 0 | 12 | 217 | 0 | 0 | 217 | 90 | 0 | 1 | 89 |
| Cambodia | 9 386 | 0 | 0 | 9 386 | 125 | 1 | 3 | 123 | 4 | 0 | 0 | 4 |
| Indonesia | 76 564 | 0 | 1 | 76 563 | 32 497 | 152 | 934 | 31 714 | 4 330 | 199 | 2 008 | 2 521 |
| Lao People's Democratic Republic | 5 928 | 0 | 0 | 5 928 | 392 | 0 | 63 | 330 | 182 | 0 | 131 | 51 |
| Malaysia | 3 119 | 2 | 0 | 3 121 | 22 000 | 116 | 5 459 | 16 657 | 5 598 | 1 160 | 3 352 | 3 406 |
| Myanmar | 37 560 | 0 | 0 | 37 560 | 4 196 | 0 | 1 476 | 2 720 | 1 056 | 0 | 275 | 781 |
| Philippines | 13 070 | 0 | 0 | 13 070 | 2 975 | 177 | 2 | 3 150 | 339 | 246 | 125 | 460 |
| Singapore | 0 | 1 | 0 | 1 | 0 | 35 | 14 | 22 | 25 | 224 | 195 | 54 |
| Thailand | 19 985 | 0 | 0 | 19 985 | 8 700 | 520 | 0 | 9 220 | 288 | 1 994 | 1 793 | 489 |
| Timor-Leste | 0 | 0 | 0 | 0 | 1 | 0 | 1 | 0 | 0 | 0 | 0 | 0 |
| Viet Nam | 21 250 | 0 | 0 | 21 250 | 5 237 | 236 | 19 | 5 454 | 2 900 | 427 | 36 | 3 290 |
| **Total Southeast Asia** | **186 874** | **3** | **2** | **186 875** | **76 339** | **1 239** | **7 971** | **69 607** | **14 812** | **4 250** | **7 916** | **11 146** |
| Bahrain | 0 | 0 | 0 | 0 | 0 | 2 | 0 | 2 | 0 | 77 | 1 | 76 |
| Cyprus | 3 | 0 | 0 | 4 | 7 | 0 | 0 | 7 | 5 | 121 | 0 | 126 |
| Iran (Islamic Republic of) | 77 | 1 | 0 | 78 | 743 | 14 | 0 | 757 | 68 | 508 | 0 | 575 |
| Iraq | 55 | 0 | 0 | 55 | 59 | 1 | 0 | 60 | 12 | 69 | 0 | 81 |
| Israel | 2 | 0 | 0 | 2 | 25 | 140 | 0 | 164 | 0 | 454 | 0 | 454 |

**NOTE:** The regional breakdown reflects geographic rather than economic or political groupings.

TABLE 4 (CONT.)

## Production, trade and consumption of roundwood and sawnwood, 2004

| Country/area | Woodfuel (1 000 m³) | | | | Industrial roundwood (1 000 m³) | | | | Sawnwood (1 000 m³) | | | |
|---|---|---|---|---|---|---|---|---|---|---|---|---|
| | Production | Imports | Exports | Consumption | Production | Imports | Exports | Consumption | Production | Imports | Exports | Consumption |
| Jordan | 253 | 0 | 0 | 253 | 4 | 7 | 2 | 10 | 0 | 256 | 7 | 249 |
| Kuwait | 0 | 0 | 0 | 0 | 0 | 7 | 0 | 7 | 0 | 129 | 0 | 129 |
| Lebanon | 82 | 0 | 0 | 82 | 7 | 38 | 1 | 45 | 9 | 248 | 1 | 256 |
| Occupied Palestinian Territory | – | – | – | – | – | – | – | – | – | – | – | – |
| Oman | 1 | 0 | 1 | 0 | 0 | 57 | 0 | 57 | 0 | 83 | 0 | 82 |
| Qatar | 0 | 0 | 0 | 0 | 0 | 34 | 1 | 33 | 0 | 33 | 0 | 32 |
| Saudi Arabia | 0 | 4 | 0 | 4 | 0 | 25 | 0 | 25 | 0 | 1 599 | 0 | 1 599 |
| Syrian Arab Republic | 18 | 0 | 0 | 19 | 40 | 17 | 6 | 51 | 9 | 387 | 2 | 394 |
| Turkey | 5 278 | 295 | 0 | 5 573 | 11 225 | 1 758 | 37 | 12 946 | 6 215 | 373 | 57 | 6 531 |
| United Arab Emirates | 0 | 0 | 0 | 0 | 0 | 160 | 3 | 156 | 0 | 484 | 12 | 472 |
| Yemen | 353 | 0 | 0 | 353 | 0 | 10 | 0 | 10 | 0 | 160 | 0 | 160 |
| **Total Western Asia** | **6 123** | **301** | **1** | **6 422** | **12 110** | **2 271** | **50** | **14 331** | **6 318** | **4 979** | **81** | **11 216** |
| **Total Asia** | **774 647** | **317** | **11** | **774 953** | **229 373** | **53 266** | **8 967** | **273 673** | **72 439** | **28 076** | **8 969** | **91 545** |
| Albania | 221 | 0 | 56 | 165 | 75 | 1 | 0 | 75 | 97 | 24 | 21 | 99 |
| Andorra | 0 | 2 | 0 | 2 | 0 | 0 | 0 | 0 | 0 | 10 | 0 | 10 |
| Austria | 3 540 | 257 | 102 | 3 695 | 12 943 | 8 812 | 935 | 20 820 | 11 133 | 1 489 | 7 396 | 5 226 |
| Belarus | 1 097 | 1 | 75 | 1 023 | 6 446 | 76 | 1 443 | 5 079 | 2 304 | 116 | 1 197 | 1 222 |
| Belgium | 600 | 19 | 22 | 598 | 4 250 | 2 879 | 1 067 | 6 062 | 1 235 | 2 249 | 1 266 | 2 218 |
| Bosnia and Herzegovina | 1 310 | 1 | 194 | 1 116 | 2 683 | 13 | 244 | 2 452 | 1 319 | 13 | 1 175 | 157 |
| Bulgaria | 2 187 | 0 | 29 | 2 158 | 2 646 | 71 | 195 | 2 522 | 332 | 7 | 273 | 66 |
| Channel Islands | – | – | – | – | – | – | – | – | – | – | – | – |
| Croatia | 954 | 2 | 151 | 805 | 2 887 | 48 | 389 | 2 546 | 582 | 338 | 355 | 565 |
| Czech Republic | 1 190 | 6 | 238 | 958 | 14 411 | 701 | 2 858 | 12 254 | 3 940 | 406 | 1 616 | 2 730 |
| Denmark | 817 | 320 | 0 | 1 136 | 810 | 501 | 309 | 1 003 | 196 | 2 251 | 134 | 2 313 |
| Estonia | 1 300 | 18 | 137 | 1 181 | 5 500 | 1 466 | 2 297 | 4 669 | 2 029 | 499 | 1 030 | 1 499 |
| Faeroe Islands | 0 | 0 | 0 | 0 | 0 | 1 | 0 | 1 | 0 | 4 | 0 | 4 |
| Finland | 4 519 | 153 | 6 | 4 666 | 49 281 | 12 961 | 525 | 61 717 | 13 544 | 404 | 8 226 | 5 722 |
| France | 2 358 | 39 | 418 | 1 979 | 31 289 | 2 175 | 3 851 | 29 614 | 9 774 | 3 829 | 1 377 | 12 226 |
| Germany | 5 847 | 120 | 32 | 5 935 | 48 657 | 2 227 | 5 589 | 45 295 | 19 538 | 5 162 | 6 212 | 18 488 |
| Gibraltar | 0 | 0 | 0 | 0 | 0 | 0 | 0 | 0 | 0 | 1 | 0 | 1 |
| Greece | 1 057 | 371 | 15 | 1 412 | 469 | 280 | 1 | 747 | 191 | 918 | 18 | 1 091 |
| Holy See | – | – | – | – | – | – | – | – | – | – | – | – |
| Hungary | 2 672 | 40 | 342 | 2 370 | 2 988 | 549 | 1 137 | 2 400 | 205 | 1 138 | 207 | 1 136 |
| Iceland | 0 | 0 | 0 | 0 | 0 | 1 | 0 | 1 | 0 | 97 | 0 | 97 |
| Ireland | 20 | 1 | 1 | 20 | 2 542 | 194 | 254 | 2 482 | 939 | 704 | 411 | 1 232 |
| Isle of Man | – | – | – | – | – | – | – | – | – | – | – | – |
| Italy | 5 814 | 803 | 0 | 6 617 | 2 883 | 4 614 | 17 | 7 481 | 1 580 | 7 661 | 157 | 9 084 |
| Latvia | 970 | 5 | 390 | 585 | 11 784 | 801 | 4 136 | 8 449 | 3 988 | 688 | 2 988 | 1 688 |

**NOTE:** The regional breakdown reflects geographic rather than economic or political groupings.

| Country/area | Woodfuel (1 000 m³) | | | | Industrial roundwood (1 000 m³) | | | | Sawnwood (1 000 m³) | | | |
|---|---|---|---|---|---|---|---|---|---|---|---|---|
| | Production | Imports | Exports | Consumption | Production | Imports | Exports | Consumption | Production | Imports | Exports | Consumption |
| Liechtenstein | 4 | 0 | 0 | 4 | 18 | 0 | 0 | 18 | – | – | – | – |
| Lithuania | 1 260 | 2 | 42 | 1 220 | 4 860 | 222 | 1 178 | 3 904 | 1 450 | 511 | 923 | 1 039 |
| Luxembourg | – | 22 | 65 | – | 264 | 420 | 255 | 429 | 133 | 64 | 51 | 147 |
| Malta | 0 | 0 | 0 | 0 | 0 | 0 | 0 | 0 | 0 | 27 | 0 | 27 |
| Monaco | – | – | – | – | – | – | – | – | – | – | – | – |
| Netherlands | 290 | 3 | 16 | 278 | 736 | 275 | 590 | 421 | 273 | 3 175 | 388 | 3 060 |
| Norway | 1 429 | 164 | 2 | 1 591 | 7 353 | 2 866 | 348 | 9 871 | 2 230 | 877 | 481 | 2 625 |
| Poland | 3 396 | 11 | 54 | 3 354 | 29 337 | 943 | 974 | 29 306 | 3 743 | 530 | 868 | 3 405 |
| Portugal | 600 | 0 | 2 | 598 | 10 953 | 364 | 1 009 | 10 308 | 1 100 | 280 | 319 | 1 061 |
| Republic of Moldova | 30 | 2 | 0 | 32 | 27 | 28 | 0 | 55 | 5 | 110 | 0 | 115 |
| Romania | 3 015 | 0 | 72 | 2 943 | 12 794 | 144 | 114 | 12 824 | 4 588 | 21 | 2 840 | 1 769 |
| Russian Federation | 47 800 | 0 | 289 | 47 511 | 130 600 | 1 004 | 41 553 | 90 051 | 21 355 | 13 | 12 621 | 8 747 |
| San Marino | – | – | – | – | – | – | – | – | – | – | – | – |
| Serbia and Montenegro | 2 097 | 2 | 5 | 2 094 | 1 423 | 11 | 33 | 1 401 | 575 | 396 | 175 | 796 |
| Slovakia | 304 | 0 | 68 | 236 | 6 936 | 246 | 1 142 | 6 040 | 1 837 | 41 | 417 | 1 461 |
| Slovenia | 725 | 11 | 79 | 657 | 1 826 | 505 | 244 | 2 086 | 461 | 224 | 411 | 274 |
| Spain | 2 055 | 18 | 101 | 1 972 | 14 235 | 2 973 | 168 | 17 040 | 3 730 | 3 326 | 80 | 6 976 |
| Sweden | 5 900 | 272 | 37 | 6 136 | 61 400 | 9 398 | 1 522 | 69 277 | 16 900 | 336 | 11 259 | 5 977 |
| Switzerland | 1 148 | 6 | 44 | 1 111 | 3 984 | 240 | 1 741 | 2 483 | 1 505 | 383 | 198 | 1 690 |
| The former Yugoslav Republic of Macedonia | 705 | 0 | 3 | 702 | 136 | 1 | 0 | 137 | 28 | 108 | 2 | 134 |
| Ukraine | 8 396 | 0 | 375 | 8 021 | 6 466 | 136 | 2 602 | 3 999 | 2 392 | 32 | 1 547 | 877 |
| United Kingdom | 231 | 7 | 151 | 88 | 8 042 | 625 | 608 | 8 058 | 2 783 | 8 653 | 371 | 11 065 |
| **Total Europe** | **115 857** | **2 680** | **3 612** | **114 968** | **503 935** | **58 771** | **79 327** | **483 379** | **138 015** | **47 114** | **67 012** | **118 117** |
| Anguilla | – | – | – | – | 0 | 0 | 0 | 0 | 0 | 0 | 0 | 0 |
| Antigua and Barbuda | – | – | – | – | 0 | 0 | 0 | 0 | 0 | 11 | 0 | 11 |
| Aruba | 0 | 0 | 0 | 0 | 0 | 1 | 0 | 1 | 0 | 16 | 0 | 16 |
| Bahamas | 0 | 0 | 0 | 0 | 17 | 67 | 4 | 80 | 1 | 103 | 2 | 102 |
| Barbados | 0 | 3 | 0 | 3 | 5 | 5 | 0 | 10 | 0 | 31 | 0 | 30 |
| Bermuda | – | – | – | – | – | – | – | – | – | – | – | – |
| British Virgin Islands | – | – | – | – | 0 | 0 | 0 | 0 | 0 | 4 | 0 | 4 |
| Cayman Islands | – | – | – | – | 0 | 2 | 0 | 2 | 0 | 14 | 0 | 14 |
| Cuba | 1 798 | 0 | 0 | 1 798 | 808 | 0 | 0 | 808 | 181 | 22 | 0 | 203 |
| Dominica | 0 | 0 | 0 | 0 | 0 | 1 | 0 | 1 | 66 | 9 | 0 | 75 |
| Dominican Republic | 556 | 0 | 0 | 556 | 6 | 13 | 0 | 19 | 0 | 267 | 0 | 267 |
| Grenada | – | – | – | – | 0 | 0 | 0 | 0 | 0 | 10 | 0 | 10 |
| Guadeloupe | 15 | 0 | 0 | 15 | 0 | 5 | 0 | 5 | 1 | 46 | 0 | 47 |
| Haiti | 1 993 | 0 | 0 | 1 993 | 239 | 1 | 0 | 240 | 14 | 19 | 0 | 33 |
| Jamaica | 570 | 0 | 0 | 570 | 282 | 3 | 0 | 285 | 66 | 38 | 0 | 104 |
| Martinique | 10 | 0 | 0 | 10 | 2 | 3 | 0 | 5 | 1 | 29 | 0 | 30 |

**NOTE:** The regional breakdown reflects geographic rather than economic or political groupings.

TABLE 4 (CONT.)

## Production, trade and consumption of roundwood and sawnwood, 2004

| Country/area | Woodfuel (1 000 m³) | | | | Industrial roundwood (1 000 m³) | | | | Sawnwood (1 000 m³) | | | |
|---|---|---|---|---|---|---|---|---|---|---|---|---|
| | Production | Imports | Exports | Consumption | Production | Imports | Exports | Consumption | Production | Imports | Exports | Consumption |
| Montserrat | – | – | – | – | – | – | – | – | 0 | 4 | 0 | 4 |
| Netherlands Antilles | 0 | 0 | 0 | 0 | 0 | 1 | 0 | 1 | 0 | 22 | 0 | 22 |
| Puerto Rico | – | – | – | – | – | – | – | – | – | – | – | – |
| Saint Kitts and Nevis | – | – | – | – | 0 | 1 | 0 | 1 | 0 | 5 | 0 | 5 |
| Saint Lucia | 0 | 0 | 0 | 0 | 0 | 7 | 0 | 7 | 0 | 15 | 0 | 15 |
| Saint Vincent and the Grenadines | 0 | 0 | 0 | 0 | 0 | 2 | 0 | 2 | 0 | 12 | 0 | 12 |
| Trinidad and Tobago | 35 | 0 | 0 | 35 | 51 | 5 | 1 | 56 | 33 | 40 | 0 | 73 |
| Turks and Caicos Islands | 0 | 0 | 0 | 0 | 0 | 0 | 0 | 0 | 0 | 4 | 0 | 4 |
| United States Virgin Islands | – | – | – | – | 0 | 0 | 0 | 0 | – | – | – | – |
| **Total Caribbean** | **4 977** | **3** | **0** | **4 980** | **1 411** | **118** | **6** | **1 524** | **363** | **718** | **3** | **1 078** |
| | | | | | | | | | | | | |
| Belize | 126 | 0 | 0 | 126 | 62 | 3 | 0 | 65 | 35 | 2 | 5 | 32 |
| Costa Rica | 3 445 | 0 | 0 | 3 445 | 1 687 | 3 | 62 | 1 628 | 812 | 29 | 3 | 838 |
| El Salvador | 4 173 | 0 | 0 | 4 173 | 682 | 2 | 10 | 674 | 16 | 43 | 0 | 59 |
| Guatemala | 15 905 | 0 | 0 | 15 905 | 419 | 13 | 10 | 421 | 366 | 7 | 41 | 332 |
| Honduras | 8 699 | 0 | 0 | 8 699 | 920 | 3 | 12 | 911 | 437 | 4 | 25 | 416 |
| Nicaragua | 5 906 | 0 | 0 | 5 906 | 93 | 0 | 29 | 64 | 45 | 1 | 39 | 6 |
| Panama | 1 219 | 0 | 0 | 1 219 | 93 | 6 | 80 | 19 | 30 | 10 | 19 | 21 |
| **Total Central America** | **39 473** | **0** | **0** | **39 473** | **3 956** | **30** | **204** | **3 782** | **1 742** | **96** | **134** | **1 704** |
| | | | | | | | | | | | | |
| Argentina | 3 972 | 0 | 0 | 3 972 | 9 706 | 7 | 39 | 9 674 | 1 388 | 86 | 234 | 1 240 |
| Bolivia | 2 228 | 0 | 0 | 2 228 | 650 | 2 | 4 | 647 | 347 | 3 | 43 | 307 |
| Brazil | 136 637 | 0 | 0 | 136 637 | 110 470 | 15 | 763 | 109 722 | 21 200 | 132 | 3 163 | 18 169 |
| Chile | 13 111 | 0 | 0 | 13 111 | 29 432 | 0 | 347 | 29 085 | 8 015 | 49 | 2 336 | 5 728 |
| Colombia | 8 469 | 0 | 0 | 8 469 | 1 993 | 2 | 23 | 1 972 | 622 | 14 | 16 | 620 |
| Ecuador | 5 427 | 0 | 0 | 5 427 | 1 211 | 0 | 47 | 1 165 | 755 | 0 | 37 | 719 |
| Falkland Islands | – | – | – | – | 0 | 0 | 0 | 0 | 0 | 0 | 0 | 0 |
| French Guiana | 95 | 0 | 0 | 95 | 60 | 1 | 2 | 59 | 15 | 1 | 4 | 12 |
| Guyana | 866 | 0 | 0 | 866 | 481 | 0 | 138 | 344 | 50 | 0 | 39 | 11 |
| Paraguay | 5 944 | 0 | 0 | 5 944 | 4 044 | 0 | 13 | 4 031 | 550 | 41 | 44 | 547 |
| Peru | 7 300 | 0 | 0 | 7 300 | 1 635 | 5 | 0 | 1 640 | 671 | 12 | 140 | 543 |
| South Georgia and the South Sandwich Islands | – | – | – | – | – | – | – | – | – | – | – | – |
| Suriname | 44 | 0 | 0 | 44 | 161 | 0 | 6 | 155 | 59 | 0 | 5 | 54 |
| Uruguay | 4 267 | 0 | 0 | 4 267 | 2 132 | 2 | 1 374 | 760 | 230 | 19 | 96 | 153 |
| Venezuela (Bolivarian Republic of) | 3 793 | 0 | 0 | 3 793 | 1 526 | 0 | 20 | 1 506 | 947 | 32 | 69 | 911 |
| **Total South America** | **192 153** | **0** | **0** | **192 153** | **163 501** | **34** | **2 776** | **160 759** | **34 849** | **391** | **6 224** | **29 015** |

**NOTE:** The regional breakdown reflects geographic rather than economic or political groupings.

| Country/area | Woodfuel (1 000 m³) | | | | Industrial roundwood (1 000 m³) | | | | Sawnwood (1 000 m³) | | | |
|---|---|---|---|---|---|---|---|---|---|---|---|---|
| | Production | Imports | Exports | Consumption | Production | Imports | Exports | Consumption | Production | Imports | Exports | Consumption |
| **Total Latin America and the Caribbean** | **236 602** | **4** | **1** | **236 605** | **168 868** | **183** | **2 986** | **166 065** | **36 954** | **1 204** | **6 361** | **31 797** |
| Canada | 2 919 | 66 | 162 | 2 823 | 197 577 | 5 961 | 3 899 | 199 639 | 60 952 | 2 994 | 41 100 | 22 847 |
| Greenland | – | – | – | – | 0 | 1 | 0 | 1 | 0 | 7 | 0 | 7 |
| Mexico | 38 269 | 0 | 7 | 38 262 | 6 913 | 262 | 6 | 7 169 | 2 962 | 2 496 | 103 | 5 355 |
| Saint Pierre and Miquelon | – | – | – | – | 0 | 0 | 0 | 0 | 0 | 2 | 0 | 2 |
| United States of America | 43 608 | 151 | 114 | 43 646 | 418 131 | 2 437 | 10 402 | 410 166 | 93 067 | 43 992 | 4 786 | 132 274 |
| **Total North America** | **84 796** | **217** | **283** | **84 731** | **622 621** | **8 660** | **14 307** | **616 975** | **156 981** | **49 491** | **45 988** | **160 484** |
| American Samoa | – | – | – | – | 0 | 0 | 0 | 0 | 0 | 1 | 0 | 1 |
| Australia | 3 092 | 0 | 0 | 3 092 | 25 685 | 2 | 1 048 | 24 639 | 4 038 | 804 | 154 | 4 688 |
| Cook Islands | – | – | – | – | 5 | 0 | 1 | 4 | 0 | 3 | 0 | 3 |
| Fiji | 37 | 0 | 0 | 37 | 346 | 4 | 7 | 342 | 84 | 1 | 35 | 49 |
| French Polynesia | – | – | – | – | 0 | 2 | 0 | 2 | 0 | 151 | 0 | 151 |
| Guam | – | – | – | – | – | – | – | – | – | – | – | – |
| Kiribati | 0 | 0 | 0 | 0 | 0 | 0 | 0 | 0 | 0 | 2 | 0 | 2 |
| Marshall Islands | – | – | – | – | – | – | – | – | 0 | 6 | 0 | 6 |
| Micronesia (Federated States of) | 0 | 0 | 0 | 0 | 0 | 0 | 0 | 0 | 0 | 7 | 0 | 7 |
| Nauru | – | – | – | – | 0 | 0 | 0 | 0 | 0 | 0 | 0 | 0 |
| New Caledonia | 0 | 0 | 0 | 0 | 5 | 4 | 3 | 5 | 3 | 30 | 1 | 33 |
| New Zealand | – | 0 | 0 | – | 19 722 | 2 | 5 240 | 14 484 | 4 369 | 41 | 1 848 | 2 562 |
| Niue | – | – | – | – | 0 | 0 | 0 | 0 | 0 | 0 | 0 | 0 |
| Northern Mariana Islands | – | – | – | – | – | – | – | – | 0 | 0 | 0 | 0 |
| Palau | – | – | – | – | 0 | 1 | 0 | 1 | 0 | 3 | 0 | 3 |
| Papua New Guinea | 5 533 | 0 | 0 | 5 533 | 2 200 | 0 | 2 012 | 188 | 60 | 0 | 16 | 44 |
| Pitcairn | – | – | – | – | – | – | – | – | 0 | 0 | 0 | 0 |
| Samoa | 70 | 0 | 0 | 70 | 61 | 14 | 1 | 74 | 21 | 14 | 0 | 35 |
| Solomon Islands | 138 | 0 | 0 | 138 | 1 020 | 0 | 1 011 | 9 | 12 | 0 | 11 | 1 |
| Tokelau | – | – | – | – | – | – | – | – | 0 | 0 | 0 | 0 |
| Tonga | 0 | 2 | 0 | 2 | 2 | 1 | 0 | 2 | 2 | 13 | 0 | 15 |
| Tuvalu | – | – | – | – | 0 | 0 | 0 | 0 | 0 | 1 | 0 | 1 |
| Vanuatu | 91 | 0 | 1 | 90 | 28 | 2 | 4 | 25 | 28 | 2 | 9 | 21 |
| Wallis and Futuna Islands | – | – | – | – | 0 | 0 | 0 | 0 | 0 | 1 | 0 | 1 |
| **Total Oceania** | **8 961** | **2** | **1** | **8 963** | **49 074** | **30** | **9 328** | **39 776** | **8 617** | **1 080** | **2 074** | **7 624** |
| **Total World** | **1 766 925** | **3 221** | **3 911** | **1 766 278** | **1 644 318** | **122 008** | **119 659** | **1 646 667** | **421 801** | **132 278** | **132 031** | **422 047** |

**SOURCE:** FAO, 2006b.

**NOTE:** The regional breakdown reflects geographic rather than economic or political groupings.

TABLE 5

## Production, trade and consumption of wood-based panels, pulp and paper, 2004

| Country/area | Wood-based panels (1 000 m³) | | | | Pulp for paper (1 000 tonnes) | | | | Paper and paperboard (1 000 tonnes) | | | |
|---|---|---|---|---|---|---|---|---|---|---|---|---|
| | Production | Imports | Exports | Consumption | Production | Imports | Exports | Consumption | Production | Imports | Exports | Consumption |
| Burundi | 0 | 1 | 0 | 1 | 0 | 0 | 0 | 0 | 0 | 3 | 0 | 3 |
| Cameroon | 88 | 0 | 51 | 37 | 0 | 0 | 0 | 0 | 0 | 60 | 0 | 60 |
| Central African Republic | 2 | 0 | 0 | 2 | 0 | 0 | 0 | 0 | 0 | 1 | 0 | 1 |
| Chad | 0 | 0 | 0 | 0 | 0 | 0 | 0 | 0 | 0 | 2 | 0 | 2 |
| Congo | 36 | 1 | 12 | 25 | 0 | 0 | 0 | 0 | 0 | 5 | 0 | 5 |
| Democratic Republic of the Congo | 3 | 1 | 1 | 3 | 0 | 0 | 0 | 0 | 3 | 9 | 1 | 11 |
| Equatorial Guinea | 27 | 1 | 26 | 2 | 0 | 0 | 0 | 0 | 0 | 0 | 0 | 0 |
| Gabon | 222 | 0 | 191 | 31 | 0 | 0 | 0 | 0 | 0 | 12 | 0 | 12 |
| Rwanda | 0 | 1 | 0 | 1 | 0 | 0 | 0 | 0 | 0 | 4 | 0 | 3 |
| Saint Helena | 0 | 0 | 0 | 0 | – | – | – | – | – | – | – | – |
| Sao Tome and Principe | 0 | 0 | 0 | 0 | 0 | 0 | 0 | 0 | 0 | 0 | 0 | 0 |
| **Total Central Africa** | **377** | **7** | **281** | **103** | **0** | **2** | **1** | **1** | **3** | **95** | **1** | **97** |
| | | | | | | | | | | | | |
| British Indian Ocean Territory | 0 | 0 | 0 | 0 | 0 | 0 | 0 | 0 | 0 | 0 | 0 | 0 |
| Comoros | 0 | 0 | 0 | 0 | 0 | 0 | 0 | 0 | 0 | 0 | 0 | 0 |
| Djibouti | 0 | 7 | 0 | 7 | 0 | 3 | 0 | 3 | 0 | 13 | 0 | 13 |
| Eritrea | 0 | 3 | 0 | 3 | 0 | 0 | 0 | 0 | 0 | 3 | 0 | 3 |
| Ethiopia | 93 | 2 | 0 | 96 | 9 | 2 | 0 | 12 | 16 | 17 | 0 | 33 |
| Kenya | 83 | 26 | 12 | 96 | 123 | 3 | 0 | 126 | 165 | 98 | 27 | 236 |
| Madagascar | 5 | 5 | 0 | 9 | 3 | 0 | 0 | 3 | 10 | 20 | 0 | 29 |
| Mauritius | 0 | 61 | 3 | 57 | 0 | 2 | 0 | 2 | 0 | 48 | 3 | 44 |
| Mayotte | – | – | – | – | – | – | – | – | – | – | – | – |
| Réunion | 0 | 24 | 0 | 23 | 0 | 0 | 0 | 0 | 0 | 15 | 0 | 15 |
| Seychelles | 0 | 1 | 0 | 1 | – | – | – | – | 0 | 0 | 0 | 0 |
| Somalia | 0 | 0 | 0 | 0 | 0 | 0 | 0 | 0 | 0 | 0 | 0 | 0 |
| Uganda | 5 | 12 | 1 | 16 | 0 | 0 | 0 | 0 | 3 | 54 | 1 | 55 |
| United Republic of Tanzania | 4 | 15 | 1 | 17 | 56 | 1 | 2 | 54 | 25 | 48 | 1 | 72 |
| **Total East Africa** | **190** | **154** | **18** | **326** | **191** | **12** | **3** | **200** | **219** | **316** | **33** | **502** |
| | | | | | | | | | | | | |
| Algeria | 48 | 146 | 0 | 194 | 2 | 17 | 0 | 19 | 41 | 399 | 2 | 438 |
| Egypt | 56 | 364 | 1 | 419 | 120 | 105 | 0 | 225 | 460 | 748 | 47 | 1 161 |
| Libyan Arab Jamahiriya | 0 | 26 | 0 | 26 | 0 | 4 | 0 | 4 | 6 | 35 | 0 | 41 |
| Mauritania | 0 | 1 | 1 | 0 | 0 | 0 | 0 | 0 | 0 | 5 | 0 | 5 |
| Morocco | 35 | 76 | 38 | 73 | 112 | 21 | 102 | 31 | 129 | 289 | 17 | 401 |
| Sudan | 2 | 26 | 0 | 27 | 0 | 0 | 1 | 0 | 3 | 34 | 1 | 37 |
| Tunisia | 104 | 84 | 22 | 165 | 10 | 97 | 29 | 78 | 121 | 215 | 52 | 283 |
| Western Sahara | – | – | – | – | – | – | – | – | – | – | – | – |
| **Total Northern Africa** | **245** | **722** | **63** | **904** | **244** | **244** | **131** | **357** | **760** | **1 725** | **120** | **2 366** |

0 = either a true zero or an insignificant value (less than half a unit)

**NOTE:** The regional breakdown reflects geographic rather than economic or political groupings.

| Country/area | Wood-based panels (1 000 m³) | | | | Pulp for paper (1 000 tonnes) | | | | Paper and paperboard (1 000 tonnes) | | | |
|---|---|---|---|---|---|---|---|---|---|---|---|---|
| | Production | Imports | Exports | Consumption | Production | Imports | Exports | Consumption | Production | Imports | Exports | Consumption |
| Angola | 11 | 5 | 0 | 16 | 15 | 0 | 0 | 15 | 0 | 19 | 0 | 19 |
| Botswana | 0 | 0 | 0 | 0 | – | – | – | – | 0 | 10 | 0 | 10 |
| Lesotho | – | – | – | – | – | – | – | – | – | – | – | – |
| Malawi | 18 | 5 | 5 | 17 | 0 | 0 | 0 | 0 | 0 | 15 | 0 | 15 |
| Mozambique | 3 | 5 | 1 | 8 | 0 | 0 | 0 | 0 | 0 | 17 | 0 | 17 |
| Namibia | – | – | – | – | – | – | – | – | 0 | 0 | 0 | 0 |
| South Africa | 1 022 | 195 | 98 | 1 119 | 1 709 | 68 | 578 | 1 199 | 3 774 | 404 | 941 | 3 237 |
| Swaziland | 8 | 0 | 0 | 8 | 191 | 0 | 191 | 0 | – | – | – | – |
| Zambia | 18 | 4 | 4 | 18 | 0 | 0 | 0 | 0 | 4 | 27 | 0 | 31 |
| Zimbabwe | 77 | 15 | 19 | 73 | 28 | 10 | 0 | 38 | 121 | 45 | 13 | 152 |
| **Total Southern Africa** | **1 156** | **230** | **127** | **1 259** | **1 943** | **79** | **769** | **1 253** | **3 899** | **536** | **955** | **3 480** |
| Benin | 0 | 1 | 0 | 1 | 0 | 0 | 0 | 0 | 0 | 8 | 0 | 7 |
| Burkina Faso | 0 | 2 | 0 | 2 | 0 | 0 | 0 | 0 | 0 | 11 | 0 | 11 |
| Cape Verde | 0 | 11 | 0 | 11 | – | – | – | – | 0 | 2 | 0 | 2 |
| Côte d'Ivoire | 340 | 0 | 202 | 138 | 0 | 0 | 0 | 0 | 0 | 71 | 2 | 69 |
| Gambia | 0 | 1 | 0 | 1 | 0 | 0 | 0 | 0 | 0 | 1 | 0 | 1 |
| Ghana | 435 | 1 | 178 | 258 | 0 | 0 | 0 | 0 | 0 | 141 | 0 | 141 |
| Guinea | 0 | 2 | 2 | 0 | 0 | 0 | 0 | 0 | 0 | 3 | 0 | 3 |
| Guinea-Bissau | 0 | 0 | 0 | 0 | – | – | – | – | 0 | 0 | 0 | 0 |
| Liberia | 250 | 0 | 0 | 250 | 0 | 0 | 0 | 0 | 0 | 2 | 0 | 1 |
| Mali | 0 | 0 | 0 | 0 | 0 | 0 | 0 | 0 | 0 | 5 | 0 | 5 |
| Niger | 0 | 0 | 0 | 0 | 0 | 9 | 0 | 9 | 0 | 4 | 0 | 4 |
| Nigeria | 95 | 42 | 0 | 136 | 23 | 17 | 0 | 40 | 19 | 297 | 2 | 315 |
| Senegal | 0 | 11 | 0 | 11 | 0 | 0 | 0 | 0 | 0 | 31 | 2 | 29 |
| Sierra Leone | 0 | 3 | 0 | 2 | 0 | 1 | 0 | 1 | 0 | 3 | 1 | 3 |
| Togo | 0 | 1 | 0 | 1 | 0 | 0 | 0 | 0 | 0 | 5 | 0 | 5 |
| **Total West Africa** | **1 120** | **76** | **384** | **812** | **23** | **29** | **0** | **52** | **19** | **584** | **8** | **595** |
| **Total Africa** | **3 088** | **1 188** | **872** | **3 404** | **2 401** | **365** | **905** | **1 862** | **4 900** | **3 257** | **1 117** | **7 040** |
| Afghanistan | 1 | 25 | 0 | 26 | 0 | 0 | 0 | 0 | 0 | 1 | 0 | 1 |
| Armenia | 2 | 37 | 0 | 39 | 0 | 0 | 0 | 0 | 2 | 12 | 0 | 14 |
| Azerbaijan | 0 | 146 | 2 | 145 | 0 | 0 | 0 | 0 | 148 | 24 | 2 | 171 |
| Georgia | 10 | 5 | 0 | 15 | 0 | 0 | 0 | 0 | 0 | 6 | 0 | 6 |
| Kazakhstan | 10 | 1 061 | 0 | 1 071 | 0 | 1 | 0 | 1 | 58 | 171 | 1 | 228 |
| Kyrgyzstan | 0 | 30 | 0 | 30 | 0 | 1 | 0 | 1 | 2 | 18 | 0 | 21 |
| Tajikistan | 0 | 9 | 0 | 9 | – | – | – | – | 0 | 1 | 0 | 1 |
| Turkmenistan | 0 | 3 | 1 | 2 | 0 | 0 | 0 | 0 | 0 | 1 | 0 | 1 |
| Uzbekistan | 0 | 198 | 2 | 196 | 0 | 1 | 1 | 0 | 11 | 38 | 3 | 47 |
| **Total Central Asia** | **24** | **1 514** | **6** | **1 532** | **0** | **3** | **1** | **2** | **222** | **274** | **6** | **489** |

**NOTE:** The regional breakdown reflects geographic rather than economic or political groupings.

TABLE 5 (CONT.)

**Production, trade and consumption of wood-based panels, pulp and paper, 2004**

| Country/area | Wood-based panels (1 000 m³) | | | | Pulp for paper (1 000 tonnes) | | | | Paper and paperboard (1 000 tonnes) | | | |
|---|---|---|---|---|---|---|---|---|---|---|---|---|
| | Production | Imports | Exports | Consumption | Production | Imports | Exports | Consumption | Production | Imports | Exports | Consumption |
| China | 44 914 | 5 491 | 5 394 | 45 011 | 16 211 | 7 679 | 53 | 23 836 | 53 463 | 10 749 | 4 193 | 60 020 |
| Democratic People's Republic of Korea | 0 | 9 | 0 | 9 | 106 | 45 | 0 | 151 | 80 | 25 | 2 | 102 |
| Japan | 5 288 | 6 462 | 40 | 11 710 | 10 703 | 2 418 | 179 | 12 942 | 29 253 | 2 274 | 1 680 | 29 847 |
| Mongolia | 2 | 4 | 1 | 5 | 0 | 0 | 0 | 0 | 0 | 5 | 0 | 5 |
| Republic of Korea | 3 860 | 2 716 | 148 | 6 428 | 545 | 2 570 | 0 | 3 115 | 10 511 | 728 | 2 996 | 8 243 |
| **Total East Asia** | **54 064** | **14 682** | **5 583** | **63 163** | **27 565** | **12 712** | **233** | **40 044** | **93 307** | **13 780** | **8 871** | **98 216** |
| Bangladesh | 9 | 13 | 0 | 22 | 65 | 32 | 0 | 97 | 58 | 239 | 1 | 297 |
| Bhutan | 32 | 0 | 23 | 9 | 0 | 1 | 0 | 0 | 0 | 1 | 1 | 0 |
| India | 2 341 | 194 | 87 | 2 448 | 3 425 | 370 | 14 | 3 781 | 4 129 | 944 | 277 | 4 795 |
| Maldives | 0 | 4 | 0 | 4 | – | – | – | – | 0 | 1 | 0 | 1 |
| Nepal | 30 | 4 | 25 | 8 | 15 | 1 | 2 | 15 | 13 | 21 | 5 | 30 |
| Pakistan | 354 | 218 | 0 | 572 | 595 | 64 | 0 | 659 | 700 | 284 | 0 | 984 |
| Sri Lanka | 22 | 55 | 60 | 16 | 21 | 3 | 0 | 24 | 25 | 257 | 1 | 280 |
| **Total South Asia** | **2 788** | **487** | **195** | **3 079** | **4 121** | **471** | **16** | **4 576** | **4 924** | **1 747** | **285** | **6 386** |
| Brunei Darussalam | 0 | 15 | 0 | 15 | 0 | 0 | 0 | 0 | 0 | 6 | 1 | 5 |
| Cambodia | 45 | 1 | 15 | 31 | 0 | 0 | 0 | 0 | 0 | 28 | 0 | 28 |
| Indonesia | 5 393 | 171 | 4 511 | 1 053 | 5 587 | 629 | 1 677 | 4 539 | 7 223 | 346 | 2 512 | 5 057 |
| Lao People's Democratic Republic | 13 | 1 | 5 | 9 | 0 | 0 | 0 | 0 | 0 | 3 | 0 | 3 |
| Malaysia | 6 963 | 468 | 6 587 | 844 | 124 | 94 | 16 | 202 | 981 | 2 046 | 347 | 2 680 |
| Myanmar | 118 | 4 | 53 | 68 | 15 | 1 | 0 | 16 | 43 | 39 | 0 | 82 |
| Philippines | 777 | 236 | 75 | 937 | 212 | 66 | 21 | 257 | 1 097 | 544 | 128 | 1 513 |
| Singapore | 355 | 314 | 147 | 522 | 0 | 7 | 0 | 7 | 87 | 699 | 163 | 623 |
| Thailand | 1 565 | 272 | 1 447 | 389 | 990 | 412 | 192 | 1 210 | 3 431 | 652 | 1 086 | 2 997 |
| Timor-Leste | 0 | 0 | 0 | 0 | 0 | 0 | 0 | 0 | 0 | 0 | 0 | 0 |
| Viet Nam | 117 | 309 | 13 | 413 | 710 | 105 | 0 | 815 | 888 | 378 | 23 | 1 242 |
| **Total Southeast Asia** | **15 346** | **1 790** | **12 854** | **4 282** | **7 638** | **1 313** | **1 905** | **7 046** | **13 749** | **4 741** | **4 260** | **14 231** |
| Bahrain | 0 | 41 | 1 | 41 | 0 | 17 | 0 | 17 | 0 | 36 | 16 | 21 |
| Cyprus | 2 | 122 | 0 | 123 | 0 | 2 | 0 | 2 | 0 | 68 | 0 | 67 |
| Iran (Islamic Republic of) | 691 | 507 | 8 | 1 190 | 345 | 72 | 0 | 417 | 411 | 688 | 7 | 1 092 |
| Iraq | 5 | 99 | 0 | 104 | 11 | 0 | 0 | 11 | 20 | 13 | 0 | 33 |
| Israel | 181 | 289 | 13 | 456 | 15 | 139 | 17 | 137 | 275 | 553 | 20 | 808 |
| Jordan | 0 | 169 | 19 | 149 | 8 | 76 | 0 | 84 | 25 | 154 | 32 | 147 |
| Kuwait | 0 | 154 | 0 | 154 | 0 | 9 | 0 | 9 | 0 | 126 | 27 | 99 |
| Lebanon | 46 | 304 | 2 | 348 | 0 | 35 | 0 | 35 | 42 | 170 | 13 | 199 |

**NOTE:** The regional breakdown reflects geographic rather than economic or political groupings.

| Country/area | Wood-based panels (1 000 m³) | | | | Pulp for paper (1 000 tonnes) | | | | Paper and paperboard (1 000 tonnes) | | | |
|---|---|---|---|---|---|---|---|---|---|---|---|---|
| | Production | Imports | Exports | Consumption | Production | Imports | Exports | Consumption | Production | Imports | Exports | Consumption |
| Occupied Palestinian Territory | – | – | – | – | – | – | – | – | – | – | – | – |
| Oman | 0 | 136 | 0 | 135 | 0 | 1 | 0 | 1 | 0 | 66 | 4 | 62 |
| Qatar | 0 | 67 | 0 | 66 | 0 | 0 | 0 | 0 | 0 | 22 | 3 | 19 |
| Saudi Arabia | 0 | 267 | 0 | 267 | 0 | 64 | 0 | 64 | 0 | 774 | 26 | 748 |
| Syrian Arab Republic | 27 | 588 | 1 | 613 | 0 | 53 | 0 | 53 | 1 | 185 | 10 | 176 |
| Turkey | 3 833 | 729 | 406 | 4 155 | 278 | 368 | 2 | 644 | 1 643 | 1 020 | 175 | 2 488 |
| United Arab Emirates | 0 | 418 | 26 | 392 | 0 | 18 | 0 | 18 | 0 | 480 | 52 | 428 |
| Yemen | 0 | 133 | 0 | 133 | 0 | 0 | 0 | 0 | 0 | 82 | 0 | 82 |
| **Total Western Asia** | **4 785** | **4 021** | **477** | **8 329** | **657** | **857** | **19** | **1 495** | **2 417** | **4 436** | **385** | **6 468** |
| **Total Asia** | **77 006** | **22 496** | **19 116** | **80 386** | **39 981** | **15 355** | **2 174** | **53 162** | **114 619** | **24 978** | **13 807** | **125 791** |
| Albania | 37 | 112 | 0 | 149 | 0 | 4 | 0 | 4 | 3 | 18 | 1 | 20 |
| Andorra | 0 | 2 | 0 | 2 | 0 | 0 | 0 | 0 | 0 | 2 | 0 | 2 |
| Austria | 3 419 | 698 | 2 689 | 1 428 | 1 698 | 674 | 259 | 2 113 | 4 852 | 1 288 | 4 128 | 2 012 |
| Belarus | 856 | 190 | 359 | 687 | 61 | 26 | 0 | 86 | 279 | 141 | 86 | 334 |
| Belgium | 2 640 | 1 878 | 3 048 | 1 470 | 531 | 959 | 696 | 793 | 2 131 | 3 025 | 2 640 | 2 516 |
| Bosnia and Herzegovina | 28 | 119 | 28 | 120 | 20 | 32 | 0 | 52 | 81 | 52 | 49 | 84 |
| Bulgaria | 520 | 137 | 280 | 377 | 102 | 13 | 60 | 55 | 171 | 160 | 52 | 279 |
| Channel Islands | – | – | – | – | – | – | – | – | – | – | – | – |
| Croatia | 103 | 250 | 68 | 285 | 109 | 1 | 43 | 67 | 464 | 183 | 136 | 511 |
| Czech Republic | 1 390 | 544 | 767 | 1 167 | 736 | 159 | 345 | 550 | 934 | 996 | 704 | 1 226 |
| Denmark | 360 | 1 621 | 166 | 1 814 | 0 | 64 | 1 | 63 | 402 | 1 268 | 279 | 1 391 |
| Estonia | 389 | 174 | 308 | 256 | 69 | 0 | 0 | 69 | 66 | 96 | 81 | 81 |
| Faeroe Islands | 0 | 1 | 0 | 1 | 0 | 0 | 0 | 0 | 0 | 2 | 0 | 1 |
| Finland | 2 024 | 281 | 1 627 | 678 | 12 614 | 168 | 2 357 | 10 425 | 14 036 | 420 | 12 708 | 1 748 |
| France | 6 146 | 1 783 | 2 987 | 4 941 | 2 426 | 2 198 | 528 | 4 096 | 10 255 | 6 302 | 5 538 | 11 019 |
| Germany | 16 350 | 5 063 | 6 962 | 14 451 | 2 502 | 4 485 | 539 | 6 448 | 20 391 | 10 465 | 11 556 | 19 300 |
| Gibraltar | 0 | 0 | 0 | 0 | 0 | 0 | 0 | 0 | 0 | 0 | 0 | 0 |
| Greece | 866 | 482 | 201 | 1 148 | 0 | 113 | 7 | 106 | 266 | 587 | 73 | 781 |
| Holy See | – | – | – | – | – | – | – | – | – | – | – | – |
| Hungary | 638 | 385 | 350 | 673 | 4 | 242 | 6 | 240 | 579 | 664 | 435 | 808 |
| Iceland | 0 | 20 | 0 | 20 | 0 | 0 | 0 | 0 | 0 | 36 | 0 | 36 |
| Ireland | 841 | 280 | 616 | 505 | 0 | 9 | 1 | 8 | 45 | 352 | 38 | 359 |
| Isle of Man | – | – | – | – | – | – | – | – | – | – | – | – |
| Italy | 5 666 | 2 161 | 1 130 | 6 697 | 657 | 3 365 | 22 | 4 000 | 9 667 | 4 893 | 3 224 | 11 337 |
| Latvia | 394 | 94 | 322 | 166 | 0 | 0 | 0 | 0 | 38 | 130 | 46 | 122 |
| Liechtenstein | – | – | – | – | – | – | – | – | – | – | – | – |
| Lithuania | 393 | 324 | 192 | 526 | 0 | 2 | 0 | 2 | 99 | 136 | 72 | 163 |
| Luxembourg | 400 | 50 | 367 | 82 | 0 | 0 | 0 | 0 | 0 | 117 | 26 | 91 |
| Malta | 0 | 35 | 1 | 34 | 0 | 0 | 0 | 0 | 0 | 33 | 1 | 32 |
| Monaco | – | – | – | – | – | – | – | – | – | – | – | – |
| Netherlands | 8 | 1 597 | 308 | 1 297 | 119 | 1 220 | 328 | 1 011 | 3 459 | 3 055 | 2 957 | 3 558 |

**NOTE:** The regional breakdown reflects geographic rather than economic or political groupings.

TABLE 5 (CONT.)

## Production, trade and consumption of wood-based panels, pulp and paper, 2004

| Country/area | Wood-based panels (1 000 m³) | | | | Pulp for paper (1 000 tonnes) | | | | Paper and paperboard (1 000 tonnes) | | | |
|---|---|---|---|---|---|---|---|---|---|---|---|---|
| | Production | Imports | Exports | Consumption | Production | Imports | Exports | Consumption | Production | Imports | Exports | Consumption |
| Norway | 589 | 247 | 266 | 570 | 2 389 | 93 | 520 | 1 961 | 2 294 | 450 | 2 004 | 741 |
| Poland | 6 491 | 1 402 | 2 292 | 5 601 | 1 029 | 472 | 41 | 1 461 | 2 635 | 1 965 | 1 262 | 3 337 |
| Portugal | 1 320 | 242 | 963 | 599 | 1 949 | 110 | 933 | 1 126 | 1 674 | 783 | 1 045 | 1 412 |
| Republic of Moldova | 10 | 25 | 0 | 34 | 0 | 0 | 0 | 0 | 0 | 27 | 8 | 19 |
| Romania | 951 | 555 | 692 | 814 | 262 | 6 | 32 | 236 | 454 | 321 | 225 | 550 |
| Russian Federation | 7 237 | 983 | 2 013 | 6 207 | 6 780 | 23 | 1 744 | 5 059 | 6 830 | 883 | 2 707 | 5 006 |
| San Marino | – | – | – | – | – | – | – | – | – | – | – | – |
| Serbia and Montenegro | 59 | 329 | 40 | 348 | 22 | 9 | 8 | 23 | 159 | 221 | 71 | 309 |
| Slovakia | 508 | 375 | 391 | 492 | 520 | 106 | 109 | 517 | 798 | 381 | 649 | 530 |
| Slovenia | 474 | 252 | 222 | 504 | 153 | 203 | 31 | 325 | 558 | 222 | 557 | 223 |
| Spain | 4 922 | 1 640 | 1 572 | 4 990 | 2 007 | 851 | 963 | 1 895 | 5 526 | 3 075 | 1 431 | 7 170 |
| Sweden | 677 | 953 | 183 | 1 447 | 12 464 | 423 | 3 388 | 9 499 | 11 589 | 628 | 10 211 | 2 006 |
| Switzerland | 897 | 586 | 878 | 606 | 237 | 505 | 80 | 663 | 1 777 | 1 128 | 1 387 | 1 518 |
| The former Yugoslav Republic of Macedonia | 0 | 71 | 2 | 70 | 0 | 0 | 0 | 0 | 16 | 28 | 6 | 38 |
| Ukraine | 1 300 | 557 | 926 | 931 | 68 | 81 | 0 | 149 | 723 | 564 | 157 | 1 130 |
| United Kingdom | 3 533 | 3 807 | 519 | 6 821 | 344 | 1 625 | 28 | 1 941 | 6 442 | 7 528 | 1 557 | 12 412 |
| **Total Europe** | **72 437** | **30 305** | **33 735** | **69 007** | **49 871** | **18 241** | **13 068** | **55 044** | **109 693** | **52 626** | **68 107** | **94 213** |
| | | | | | | | | | | | | |
| Anguilla | 0 | 0 | 0 | 0 | – | – | – | – | 0 | 0 | 0 | 0 |
| Antigua and Barbuda | 0 | 4 | 0 | 4 | – | – | – | – | 0 | 0 | 0 | 0 |
| Aruba | 0 | 6 | 0 | 6 | 0 | 0 | 0 | 0 | 0 | 1 | 0 | 1 |
| Bahamas | 0 | 16 | 0 | 16 | 0 | 0 | 0 | 0 | – | – | – | – |
| Barbados | 0 | 16 | 0 | 16 | 0 | 1 | 0 | 1 | 0 | 17 | 0 | 17 |
| Bermuda | – | – | – | – | 0 | 0 | 0 | 0 | – | – | – | – |
| British Virgin Islands | 0 | 1 | 0 | 1 | 0 | 0 | 0 | 0 | 0 | 0 | 0 | 0 |
| Cayman Islands | 0 | 5 | 0 | 5 | 0 | 0 | 0 | 0 | 0 | 1 | 0 | 1 |
| Cuba | 149 | 33 | 1 | 181 | 1 | 6 | 0 | 7 | 27 | 68 | 0 | 95 |
| Dominica | 0 | 2 | 1 | 2 | 0 | 0 | 0 | 0 | 0 | 1 | 0 | 1 |
| Dominican Republic | 0 | 31 | 0 | 31 | 0 | 1 | 0 | 1 | 130 | 182 | 1 | 311 |
| Grenada | 0 | 4 | 0 | 4 | – | – | – | – | 0 | 0 | 0 | 0 |
| Guadeloupe | 0 | 23 | 0 | 23 | 0 | 0 | 0 | 0 | 0 | 6 | 0 | 6 |
| Haiti | 0 | 2 | 0 | 2 | 0 | 0 | 0 | 0 | 0 | 9 | 0 | 9 |
| Jamaica | 0 | 70 | 0 | 70 | 0 | 0 | 0 | 0 | 0 | 35 | 0 | 35 |
| Martinique | 0 | 7 | 0 | 7 | 0 | 0 | 0 | 0 | 0 | 5 | 0 | 5 |
| Montserrat | 0 | 0 | 0 | 0 | – | – | – | – | – | – | – | – |
| Netherlands Antilles | 0 | 10 | 0 | 10 | 0 | 0 | 0 | 0 | – | – | – | – |
| Puerto Rico | – | – | – | – | – | – | – | – | – | – | – | – |
| Saint Kitts and Nevis | 0 | 1 | 0 | 1 | – | – | – | – | 0 | 0 | 0 | 0 |
| Saint Lucia | 0 | 7 | 0 | 7 | 0 | 0 | 0 | 0 | 0 | 10 | 0 | 10 |

**NOTE:** The regional breakdown reflects geographic rather than economic or political groupings.

| Country/area | Wood-based panels (1 000 m³) | | | | Pulp for paper (1 000 tonnes) | | | | Paper and paperboard (1 000 tonnes) | | | |
|---|---|---|---|---|---|---|---|---|---|---|---|---|
| | Production | Imports | Exports | Consumption | Production | Imports | Exports | Consumption | Production | Imports | Exports | Consumption |
| Saint Vincent and the Grenadines | 0 | 2 | 0 | 2 | – | – | – | – | 0 | 3 | 0 | 3 |
| Trinidad and Tobago | 0 | 44 | 0 | 44 | 0 | 4 | 0 | 4 | 0 | 100 | 1 | 99 |
| Turks and Caicos Islands | 0 | 1 | 0 | 1 | – | – | – | – | 0 | 0 | 0 | 0 |
| United States Virgin Islands | – | – | – | – | – | – | – | – | – | – | – | – |
| **Total Caribbean** | **149** | **287** | **2** | **433** | **1** | **12** | **0** | **13** | **157** | **440** | **3** | **595** |
| Belize | 0 | 3 | 1 | 2 | 0 | 2 | 1 | 1 | 0 | 4 | 0 | 3 |
| Costa Rica | 65 | 50 | 33 | 82 | 10 | 33 | 0 | 42 | 20 | 392 | 22 | 390 |
| El Salvador | 0 | 41 | 0 | 40 | 0 | 6 | 1 | 5 | 56 | 180 | 15 | 221 |
| Guatemala | 43 | 63 | 13 | 93 | 0 | 1 | 0 | 1 | 31 | 293 | 16 | 308 |
| Honduras | 9 | 25 | 0 | 34 | 7 | 1 | 0 | 8 | 95 | 152 | 2 | 245 |
| Nicaragua | 8 | 13 | 4 | 18 | 0 | 0 | 0 | 0 | 0 | 37 | 1 | 36 |
| Panama | 7 | 27 | 0 | 34 | 0 | 2 | 0 | 2 | 0 | 98 | 28 | 70 |
| **Total Central America** | **133** | **223** | **52** | **303** | **17** | **44** | **2** | **60** | **202** | **1 155** | **84** | **1 273** |
| Argentina | 1 112 | 48 | 691 | 470 | 894 | 150 | 291 | 754 | 1 521 | 684 | 192 | 2 013 |
| Bolivia | 12 | 9 | 12 | 9 | 0 | 0 | 0 | 0 | 0 | 45 | 0 | 45 |
| Brazil | 6 283 | 255 | 3 707 | 2 831 | 9 529 | 397 | 4 026 | 5 900 | 8 221 | 654 | 1 651 | 7 224 |
| Chile | 1 927 | 36 | 1 008 | 954 | 3 338 | 17 | 2 545 | 810 | 1 170 | 452 | 547 | 1 075 |
| Colombia | 225 | 176 | 70 | 331 | 381 | 135 | 1 | 515 | 899 | 413 | 179 | 1 133 |
| Ecuador | 261 | 67 | 121 | 207 | 2 | 24 | 0 | 26 | 100 | 232 | 21 | 311 |
| Falkland Islands | 0 | 0 | 0 | 0 | – | – | – | – | 0 | 0 | 0 | 0 |
| French Guiana | 0 | 3 | 0 | 3 | 0 | 0 | 0 | 0 | 0 | 0 | 0 | 0 |
| Guyana | 54 | 2 | 53 | 3 | 0 | 0 | 0 | 0 | 0 | 4 | 0 | 4 |
| Paraguay | 161 | 5 | 31 | 135 | 0 | 0 | 0 | 0 | 13 | 75 | 3 | 85 |
| Peru | 97 | 86 | 32 | 151 | 17 | 41 | 0 | 58 | 91 | 291 | 17 | 364 |
| South Georgia and the South Sandwich Islands | – | – | – | – | – | – | – | – | – | – | – | – |
| Suriname | 1 | 8 | 2 | 7 | 0 | 0 | 0 | 0 | 0 | 3 | 0 | 3 |
| Uruguay | 6 | 24 | 0 | 30 | 41 | 15 | 1 | 55 | 96 | 57 | 78 | 75 |
| Venezuela (Bolivarian Republic of) | 151 | 40 | 127 | 64 | 142 | 119 | 0 | 261 | 723 | 288 | 45 | 966 |
| **Total South America** | **10 290** | **760** | **5 855** | **5 196** | **14 344** | **898** | **6 863** | **8 379** | **12 833** | **3 197** | **2 732** | **13 298** |
| **Total Latin America and the Caribbean** | **10 572** | **1 269** | **5 909** | **5 932** | **14 362** | **955** | **6 865** | **8 451** | **13 193** | **4 792** | **2 819** | **15 166** |
| Canada | 16 617 | 1 612 | 13 383 | 4 846 | 26 222 | 281 | 11 380 | 15 123 | 20 599 | 3 709 | 16 122 | 8 186 |
| Greenland | 0 | 5 | 0 | 5 | 0 | 0 | 0 | 0 | 0 | 1 | 0 | 1 |
| Mexico | 430 | 1 869 | 249 | 2 049 | 375 | 837 | 37 | 1 175 | 4 391 | 2 075 | 211 | 6 255 |
| Saint Pierre and Miquelon | 0 | 1 | 0 | 0 | – | – | – | – | 0 | 0 | 0 | 0 |
| United States of America | 44 514 | 21 077 | 2 940 | 62 651 | 53 817 | 6 096 | 5 450 | 54 463 | 82 084 | 17 513 | 9 033 | 90 565 |
| **Total North America** | **61 561** | **24 563** | **16 572** | **69 552** | **80 414** | **7 214** | **16 867** | **70 761** | **107 074** | **23 298** | **25 366** | **105 006** |

**NOTE:** The regional breakdown reflects geographic rather than economic or political groupings.

TABLE 5 (CONT.)

**Production, trade and consumption of wood-based panels, pulp and paper, 2004**

| Country/area | Wood-based panels (1 000 m³) | | | | Pulp for paper (1 000 tonnes) | | | | Paper and paperboard (1 000 tonnes) | | | |
|---|---|---|---|---|---|---|---|---|---|---|---|---|
| | Production | Imports | Exports | Consumption | Production | Imports | Exports | Consumption | Production | Imports | Exports | Consumption |
| American Samoa | 0 | 0 | 0 | 0 | – | – | – | – | 0 | 0 | 0 | 0 |
| Australia | 2 083 | 406 | 566 | 1 923 | 1 107 | 368 | 4 | 1 471 | 3 097 | 1 564 | 776 | 3 885 |
| Cook Islands | 0 | 2 | 0 | 2 | – | – | – | – | – | – | – | – |
| Fiji | 16 | 17 | 9 | 24 | 0 | 0 | 0 | 0 | 0 | 33 | 2 | 31 |
| French Polynesia | 0 | 19 | 0 | 19 | 0 | 0 | 0 | 0 | 0 | 15 | 0 | 15 |
| Guam | – | – | – | – | – | – | – | – | – | – | – | – |
| Kiribati | 0 | 0 | 0 | 0 | – | – | – | – | 0 | 0 | 0 | 0 |
| Marshall Islands | 0 | 3 | 0 | 3 | – | – | – | – | 0 | 0 | 0 | 0 |
| Micronesia (Federated States of) | 0 | 1 | 0 | 1 | – | – | – | – | 0 | 0 | 0 | 0 |
| Nauru | 0 | 0 | 0 | 0 | – | – | – | – | – | – | – | – |
| New Caledonia | 0 | 22 | 3 | 19 | 0 | 0 | 0 | 0 | 0 | 15 | 8 | 7 |
| New Zealand | 2 219 | 23 | 1 064 | 1 178 | 1 596 | 6 | 861 | 741 | 920 | 455 | 622 | 753 |
| Niue | 0 | 0 | 0 | 0 | 0 | 0 | 0 | 0 | 0 | 0 | 0 | 0 |
| Northern Mariana Islands | 0 | 0 | 0 | 0 | – | – | – | – | 0 | 0 | 0 | 0 |
| Palau | 0 | 1 | 0 | 1 | – | – | – | – | 0 | 0 | 0 | 0 |
| Papua New Guinea | 69 | 0 | 65 | 4 | – | – | – | – | 0 | 16 | 0 | 16 |
| Pitcairn | 0 | 0 | 0 | 0 | – | – | – | 0 | 0 | 0 | 0 | 0 |
| Samoa | 0 | 5 | 0 | 5 | 0 | 0 | 0 | 0 | 0 | 3 | 0 | 3 |
| Solomon Islands | 0 | 0 | 0 | 0 | – | – | – | – | 0 | 0 | 0 | 0 |
| Tokelau | – | – | – | – | – | – | – | – | 0 | 0 | 0 | 0 |
| Tonga | 0 | 1 | 0 | 1 | – | – | – | – | 0 | 1 | 0 | 1 |
| Tuvalu | 0 | 0 | 0 | 0 | – | – | – | – | 0 | 0 | 0 | 0 |
| Vanuatu | 0 | 0 | 0 | 0 | 0 | 1 | 0 | 1 | 0 | 1 | 0 | 1 |
| Wallis and Futuna Islands | 0 | 0 | 0 | 0 | – | – | – | – | – | – | – | – |
| **Total Oceania** | **4 387** | **502** | **1 707** | **3 182** | **2 703** | **375** | **865** | **2 213** | **4 017** | **2 103** | **1 408** | **4 712** |
| **Total World** | **229 051** | **80 323** | **77 910** | **231 464** | **189 732** | **42 505** | **40 744** | **191 493** | **353 496** | **111 055** | **112 624** | **351 928** |

**SOURCE:** FAO, 2006b.

**NOTE:** The regional breakdown reflects geographic rather than economic or political groupings.

TABLE 6

## Status of ratification of international conventions and agreements as of 1 January 2007

| Country/territory | CBD | UNFCCC | Kyoto Protocol | CCD | CITES | Ramsar Convention | World Heritage Convention |
|---|---|---|---|---|---|---|---|
| **Africa** | | | | | | | |
| Algeria | X | X | X | X | X | X | X |
| Angola | X | X | | X | | | X |
| Benin | X | X | X | X | X | X | X |
| Botswana | X | X | X | X | X | X | X |
| Burkina Faso | X | X | X | X | X | X | X |
| Burundi | X | X | X | X | X | X | X |
| Cameroon | X | X | X | X | X | X | X |
| Cape Verde | X | X | X | X | X | X | X |
| Central African Republic | X | X | | X | X | X | X |
| Chad | X | X | | X | X | X | X |
| Comoros | X | X | | X | X | X | X |
| Congo | X | X | | X | X | X | X |
| Cote d'Ivoire | X | X | | X | X | X | X |
| Dem. Republic of the Congo | X | X | X | X | X | X | X |
| Djibouti | X | X | X | X | X | X | |
| Egypt | X | X | X | X | X | X | X |
| Equatorial Guinea | X | X | X | X | X | X | |
| Eritrea | X | X | X | X | X | | X |
| Ethiopia | X | X | X | X | X | | X |
| Gabon | X | X | X | X | X | X | X |
| Gambia | X | X | X | X | X | X | X |
| Ghana | X | X | X | X | X | X | X |
| Guinea | X | X | X | X | X | X | X |
| Guinea-Bissau | X | X | X | X | X | X | X |
| Kenya | X | X | X | X | X | X | X |
| Lesotho | X | X | X | X | X | X | X |
| Liberia | X | X | X | X | X | X | X |
| Libyan Arab Jamahiriya | X | X | X | X | X | X | X |
| Madagascar | X | X | X | X | X | X | X |
| Malawi | X | X | X | X | X | X | X |
| Mali | X | X | X | X | X | X | X |
| Mauritania | X | X | X | X | X | X | X |
| Mauritius | X | X | X | X | X | X | X |
| Morocco | X | X | X | X | X | X | X |
| Mozambique | X | X | X | X | X | X | X |
| Namibia | X | X | X | X | X | X | X |
| Niger | X | X | X | X | X | X | X |
| Nigeria | X | X | X | X | X | X | X |
| Rwanda | X | X | X | X | X | X | X |
| Sao Tome and Principe | X | X | | X | X | X | X |
| Senegal | X | X | X | X | X | X | X |
| Seychelles | X | X | X | X | X | X | X |
| Sierra Leone | X | X | X | X | X | X | X |
| Somalia | | | | X | X | | |
| South Africa | X | X | X | X | X | X | X |
| Sudan | X | X | X | X | X | X | X |
| Swaziland | X | X | X | X | X | | X |
| Togo | X | X | X | X | X | X | X |
| Tunisia | X | X | X | X | X | X | X |
| Uganda | X | X | X | X | X | X | X |
| United Republic of Tanzania | X | X | X | X | X | X | X |
| Zambia | X | X | X | X | X | X | X |
| Zimbabwe | X | X | | X | X | | X |

**NOTE:** The regional breakdown reflects geographic rather than economic or political groupings.

TABLE 6 (CONT.)

**Status of ratification of international conventions and agreements as of 1 January 2007**

| Country/territory | CBD | UNFCCC | Kyoto Protocol | CCD | CITES | Ramsar Convention | World Heritage Convention |
|---|---|---|---|---|---|---|---|
| **Asia** | | | | | | | |
| Afghanistan | X | X | | X | X | | X |
| Armenia | X | X | X | X | | X | X |
| Azerbaijan | X | X | X | X | X | X | X |
| Bahrain | X | X | X | X | | X | X |
| Bangladesh | X | X | X | X | X | X | X |
| Bhutan | X | X | X | X | X | | X |
| Brunei Darussalam | | | | X | X | | |
| Cambodia | X | X | X | X | X | X | X |
| China | X | X | X | X | X | X | X |
| Cyprus | X | X | X | X | X | X | X |
| Dem. People's Rep. of Korea | X | X | X | X | | | X |
| Georgia | X | X | X | X | X | X | X |
| India | X | X | X | X | X | X | X |
| Indonesia | X | X | X | X | X | X | X |
| Iran, Islamic Rep. of | X | X | X | X | X | X | X |
| Iraq | | | | | | | X |
| Israel | X | X | X | X | X | X | X |
| Japan | X | X | X | X | X | X | X |
| Jordan | X | X | X | X | X | X | X |
| Kazakhstan | X | X | | X | X | | X |
| Kuwait | X | X | X | X | X | | X |
| Kyrgyzstan | X | X | X | X | | X | X |
| Lao People's Dem. Rep. | X | X | X | X | X | | X |
| Lebanon | X | X | X | X | | X | X |
| Malaysia | X | X | X | X | X | X | X |
| Maldives | X | X | X | X | | | X |
| Mongolia | X | X | X | X | X | X | X |
| Myanmar | X | X | X | X | X | X | X |
| Nepal | X | X | X | X | X | X | X |
| Oman | X | X | X | X | | | X |
| Pakistan | X | X | X | X | X | X | X |
| Philippines | X | X | X | X | X | X | X |
| Qatar | X | X | X | X | X | | X |
| Republic of Korea | X | X | X | X | X | X | X |
| Saudi Arabia | X | X | X | X | X | | X |
| Singapore | X | X | X | X | X | | |
| Sri Lanka | X | X | X | X | X | X | X |
| Syrian Arab Republic | X | X | X | X | X | X | X |
| Tajikistan | X | X | | X | | X | X |
| Thailand | X | X | X | X | X | X | X |
| Timor-Leste | X | X | | X | | | |
| Turkey | X | X | | X | X | X | X |
| Turkmenistan | X | X | X | X | | | X |
| United Arab Emirates | X | X | X | X | X | | X |
| Uzbekistan | X | X | X | X | X | X | X |
| Viet Nam | X | X | X | X | X | X | X |
| Yemen | X | X | X | X | X | | X |
| **Europe** | | | | | | | |
| Albania | X | X | X | X | X | X | X |
| Andorra | | | | X | | | X |
| Austria | X | X | X | X | X | X | X |
| Belarus | X | X | X | X | X | X | X |
| Belgium | X | X | X | X | X | X | X |

**NOTE:** The regional breakdown reflects geographic rather than economic or political groupings.

| Country/territory | CBD | UNFCCC | Kyoto Protocol | CCD | CITES | Ramsar Convention | World Heritage Convention |
|---|---|---|---|---|---|---|---|
| Bosnia and Herzegovina | X | X |  | X |  | X | X |
| Bulgaria | X | X | X | X | X | X | X |
| Croatia | X | X |  | X | X | X | X |
| Czech Republic | X | X | X | X | X | X | X |
| Denmark | X | X | X | X | X | X | X |
| Estonia | X | X | X |  | X | X | X |
| Finland | X | X | X | X | X | X | X |
| France | X | X | X | X | X | X | X |
| Germany | X | X | X | X | X | X | X |
| Greece | X | X | X | X | X | X | X |
| Hungary | X | X | X | X | X | X | X |
| Iceland | X | X | X | X | X | X | X |
| Ireland | X | X | X | X | X | X | X |
| Italy | X | X | X | X | X | X | X |
| Latvia | X | X | X | X | X | X | X |
| Liechtenstein | X | X | X | X | X | X |  |
| Lithuania | X | X | X | X | X | X | X |
| Luxembourg | X | X | X | X | X | X | X |
| Malta | X | X | X | X | X | X | X |
| Monaco | X | X | X | X | X | X | X |
| Netherlands | X | X | X | X | X | X | X |
| Norway | X | X | X | X | X | X | X |
| Poland | X | X | X | X | X | X | X |
| Portugal | X | X | X | X | X | X | X |
| Republic of Moldova | X | X | X | X | X | X | X |
| Romania | X | X | X | X | X | X | X |
| Russian Federation | X | X | X | X | X | X | X |
| San Marino | X | X |  | X | X |  | X |
| Serbia and Montenegro | X | X |  |  | X | X | X |
| Slovakia | X | X | X | X | X | X | X |
| Slovenia | X | X | X | X | X | X | X |
| Spain | X | X | X | X | X | X | X |
| Sweden | X | X | X | X | X | X | X |
| Switzerland | X | X | X | X | X | X | X |
| The FYR of Macedonia | X | X | X | X | X | X | X |
| Ukraine | X | X | X | X | X | X | X |
| United Kingdom | X | X | X | X | X | X | X |
| **North and Central America** |  |  |  |  |  |  |  |
| Antigua and Barbuda | X | X | X | X | X | X | X |
| Bahamas | X | X | X | X | X | X |  |
| Barbados | X | X | X | X | X | X | X |
| Belize | X | X | X | X | X | X | X |
| Canada | X | X | X | X | X | X | X |
| Cayman Islands |  |  |  |  |  |  |  |
| Costa Rica | X | X | X | X | X | X | X |
| Cuba | X | X | X | X | X | X | X |
| Dominica | X | X | X | X | X |  | X |
| Dominican Republic | X | X | X | X | X | X | X |
| El Salvador | X | X | X | X | X | X | X |
| Greenland |  |  |  |  |  |  |  |
| Grenada | X | X | X | X | X |  | X |
| Guatemala | X | X | X | X | X | X | X |
| Haiti | X | X | X | X |  |  | X |
| Honduras | X | X | X | X | X | X | X |
| Jamaica | X | X | X | X | X | X | X |

**NOTE:** The regional breakdown reflects geographic rather than economic or political groupings.

TABLE 6 (CONT.)
**Status of ratification of international conventions and agreements as of 1 January 2007**

| Country/territory | CBD | UNFCCC | Kyoto Protocol | CCD | CITES | Ramsar Convention | World Heritage Convention |
|---|---|---|---|---|---|---|---|
| Mexico | X | X | X | X | X | X | X |
| Nicaragua | X | X | X | X | X | X | X |
| Panama | X | X | X | X | X | X | X |
| Saint Kitts and Nevis | X | X | | X | X | | X |
| Saint Lucia | X | X | X | X | X | X | X |
| Saint Vincent and Grenadines | X | X | X | X | X | | X |
| Trinidad and Tobago | X | X | X | X | X | X | X |
| United States | | X | | X | X | X | X |
| United States Virgin Islands | | | | | | | |
| **Oceania** | | | | | | | |
| American Samoa | | | | | | | |
| Australia | X | X | | X | X | X | X |
| Cook Islands | X | X | X | X | | | |
| Fiji | X | X | X | X | X | X | X |
| French Polynesia | | | | | | | |
| Guam | | | | | | | |
| Kiribati | X | X | X | X | | | X |
| Marshall Islands | X | X | X | X | | X | X |
| Micronesia | X | X | X | X | | | X |
| Nauru | X | X | X | X | | | |
| New Caledonia | | | | | | | |
| New Zealand | X | X | X | X | X | X | X |
| Niue | X | X | X | X | | | X |
| Northern Mariana Islands | | | | | | | |
| Palau | X | X | X | X | X | X | X |
| Papua New Guinea | X | X | X | X | X | X | X |
| Samoa | X | X | X | X | X | X | X |
| Solomon Islands | X | X | X | X | | | X |
| Tonga | X | X | | X | | | X |
| Tuvalu | X | X | X | X | | | |
| Vanuatu | X | X | X | X | X | | X |
| **South America** | | | | | | | |
| Argentina | X | X | X | X | X | X | X |
| Bolivia | X | X | X | X | X | X | X |
| Brazil | X | X | X | X | X | X | X |
| Chile | X | X | X | X | X | X | X |
| Colombia | X | X | X | X | X | X | X |
| Ecuador | X | X | X | X | X | X | X |
| Guyana | X | X | X | X | X | | X |
| Paraguay | X | X | X | X | X | X | X |
| Peru | X | X | X | X | X | X | X |
| Suriname | X | X | X | X | X | X | X |
| Uruguay | X | X | X | X | X | X | X |
| Venezuela, Bolivarian Rep. of | X | X | X | X | X | X | X |

**NOTE:** The regional breakdown reflects geographic rather than economic or political groupings.